Lumos Test Mastery
Grade 3 Math Practice

ONLINE REGISTRATION REQUIRED

Visit the link below for online registration

lumoslearning.com/a/tedbooks

Access Code: TMG3MATH-32891-P

Why is Registration Mandatory?

2 Lumos Diagnostic Tests

Personalized Study Plan

Automated Grading

Answer Key & Explanations

INTRODUCTION

Lumos Test Mastery is an innovative blended test prep program that combines both print and online resources to help students excel in high-stakes state assessments and boost their grade-level skills.

The printed workbook features a wide variety of standards-aligned practice questions, giving students the opportunity to practice at their own pace and focus on the areas where they need the most help. These workbooks serve as invaluable tools for mastering critical state assessment skills, reinforcing concepts and abilities, bridging learning gaps, and much more.

Complementing this, the online program provides access to two comprehensive Lumos diagnostic tests that feature an array of technology-enhanced question types to meticulously evaluate student proficiency across essential standards. Based on the student's performance on the diagnostic tests, the online program crafts personalized study plans to deliver targeted remedial practice. Furthermore, the program grants access to an expansive question bank, enabling educators to craft their own questions and assessments that mirror the state tests.

Students can conveniently answer workbook questions using either printed or virtual bubble sheets, leading to swift, automated grading with instant feedback. This eliminates the need for manual grading of student responses, saving valuable time for parents and teachers.

START A
Diagnostic Test-1
D Diagnostic Test-2
FINISH
E B
Targeted Practice
State Test
Study Plan
C F

A Students take the first diagnostic test to evaluate their proficiency across essential standards.

B Based on their performance, the program creates a personalized study plan for targeted practice.

C Students complete the targeted remedial practice outlined in the study plan.

D Students take the second diagnostic test to gauge their mastery and progress.

E The program generates a second study plan, uncovering areas that demand further practice.

F Students complete the targeted practice to gain comprehensive mastery needed to ensure success on the state test.

How to Use the Lumos Program

<u>GETTING STARTED</u>

Parents & Caregivers: Register Online using the Access Code or by scanning the QR Code provided on the first page of this book. Select the *User Type as 'Student/Parent'* to get the login details for the Student and Parent Accounts in your registered email id.

Teachers: Register Online using the Access Code or by scanning the QR Code provided on the first page of this book. Select the *User Type as Teacher* to get the login details for the Teacher Account in your registered email id. Using the *My Student* option on the menu, add a student. Provide the login details to your student for online access. Please note that access for 1 student is included in the program. For additional student licenses and school purchases, contact us at **support@lumoslearning.com**

<u>TAKING THE DIAGNOSTIC TEST:</u> Login to the student account. Click on the *Bell Icon* on the top right-hand corner or *Start Test* button on the home page. The test is also available on the table of contents in the student account.

<u>GETTING THE STUDY PLAN:</u> On completing the diagnostic test, a personalized study plan is generated which outlines the lessons to be practiced in this printed workbook. To access the study plan, click on the *STUDY PLAN* button on the online table of contents. The student must complete the study plan by practicing the lessons in this workbook.

<u>COMPLETING LESSONS IN THIS WORKBOOK:</u> Using the table of contents in this book, goto the lessons mentioned in the study plan. You will find a *Printed Bubble Sheet* before every lesson which may be used by the student to answer the questions. Instructions on *How to Use the Printed Bubble Sheet* are provided at the end of this book. Alternatively, the student may also answer the questions using the *Virtual Bubble Sheet* in the student account.

<u>GRADING ANSWERS:</u> On submitting the answers, the progam uses the *Automated Grading* feature to get the results. For grading open-ended questions, login to your parent/teacher account and grade them manually.

<u>ANSWERS & EXPLANATIONS:</u> Answers and detailed explanations are provided in the student results page. You can also login to your parent/teacher account and click on the *Preview* icon under *My Lessons* to view the answer explanations.

<u>MONITORING STUDENT PROGRESS:</u> You can view the performance reports of the student in the *Reports* section of the parent/teacher account.

<u>COMPLETING THE PROGRAM:</u> After completing the first study plan, take the second diagnostic test. Finish the study plan before the actual state test.

Table of Contents

Chapter 1

Operations and Algebraic Thinking

Test ID	3M001
Test Name	Understanding Multiplication
Student Name	
Date	

1	Ⓐ Ⓑ Ⓒ Ⓓ	6	Ⓐ Ⓑ Ⓒ Ⓓ	11	Ⓐ Ⓑ Ⓒ Ⓓ	16	Answer in the space provided below.
2	Ⓐ Ⓑ Ⓒ Ⓓ	7	Ⓐ Ⓑ Ⓒ Ⓓ	12	Ⓐ Ⓑ Ⓒ Ⓓ	17	Answer in the space provided below.
3	Ⓐ Ⓑ Ⓒ Ⓓ	8	Ⓐ Ⓑ Ⓒ Ⓓ	13	Ⓐ Ⓑ Ⓒ Ⓓ	18	Answer in the space provided below.
4	Ⓐ Ⓑ Ⓒ Ⓓ	9	Ⓐ Ⓑ Ⓒ Ⓓ	14	Ⓐ Ⓑ Ⓒ Ⓓ	19	Answer in the space provided below.
5	Ⓐ Ⓑ Ⓒ Ⓓ	10	Ⓐ Ⓑ Ⓒ Ⓓ	15	Ⓐ Ⓑ Ⓒ Ⓓ	20	Answer in the space provided below.
						21	Answer in the space provided below.

16

17

	Column A 3+3+3+3+3+3	Column B 3+3+3+3+3+3+3+3	Column C 3+3+3
3 x 8	○	○	○
3 x 3	○	○	○
3 x 6	○	○	○

18

19

20

21

Number of lions	5	6	9		
Total number of legs	20			32	16

Date of Completion:_______________ Score:_______________ 

Lesson 1: Understanding Multiplication

1. Which multiplication fact is being modeled below?

□□□□□□□□□□
□□□□□□□□□□
□□□□□□□□□□

Ⓐ 3 x 10 = 30
Ⓑ 4 x 10 = 40
Ⓒ 4 x 9 = 36
Ⓓ 3 x 9 = 27

2. Which numerical expression describes this array?

○○○○○
○○○○○
○○○○○
○○○○○

Ⓐ 4 + 5
Ⓑ 5 + 4
Ⓒ 4 x 5
Ⓓ 4 x 4

3. Which number sentence describes this array?

○○○○○○○
○○○○○○○
○○○○○○○
○○○○○○○

Ⓐ 8 x 4 = 32
Ⓑ 7 + 5 = 12
Ⓒ 5 x 7 = 35
Ⓓ 4 x 7 = 28

4. **Which number sentence describes this array?**

Ⓐ **2 x 12 = 24**
Ⓑ **2 + 12 = 14**
Ⓒ **12 + 2 = 24**
Ⓓ **10 x 2 = 20**

5. **Identify the multiplication sentence for the picture below:**

Ⓐ **4 x 4 = 16**
Ⓑ **4 x 3 = 12**
Ⓒ **3 x 4 = 12**
Ⓓ **4 x 2 = 8**

6. **What multiplication fact does this picture model?**

Ⓐ **4 x 6 = 24**
Ⓑ **4 x 7 = 28**
Ⓒ **6 x 3 = 18**
Ⓓ **7 x 4 = 28**

7. **Identify the multiplication sentence for the picture below:**

Ⓐ 7 x 2 = 14
Ⓑ 7 x 3 = 21
Ⓒ 7 x 4 = 28
Ⓓ 6 x 3 = 18

8. **Identify the multiplication sentence for the picture below:**

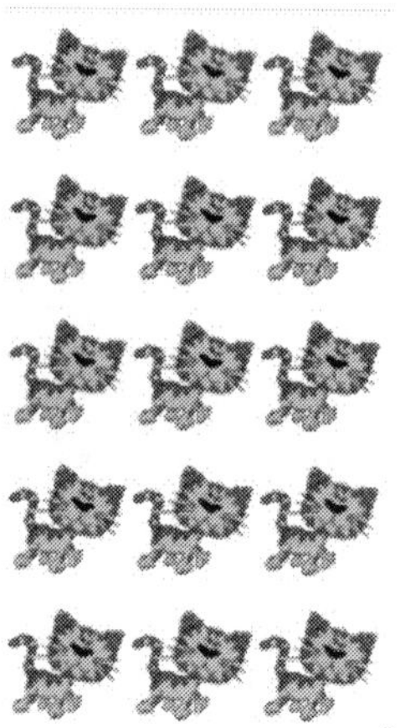

Ⓐ 4 x 4 = 16
Ⓑ 3 x 6 = 18
Ⓒ 3 x 4 = 12
Ⓓ 3 x 5 = 15

9. Identify the multiplication sentence for the picture below:

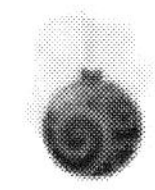  

Ⓐ 3 x 2 = 6
Ⓑ 3 x 3 = 9
Ⓒ 4 x 2 = 8
Ⓓ 3 x 1 = 3

10. Identify the multiplication sentence for the picture below:

Ⓐ 3 x 5 = 15
Ⓑ 4 x 4 = 16
Ⓒ 5 x 4 = 20
Ⓓ 7 x 4 = 28

11. Identify the multiplication sentence for the picture below:

Ⓐ 2 x 5 = 10
Ⓑ 4 x 2 = 8
Ⓒ 4 x 1 = 4
Ⓓ 4 + 2 = 6

12. Identify the multiplication sentence for the picture below:

Ⓐ 6 x 7 = 42
Ⓑ 6 x 8 = 48
Ⓒ 8 x 9 = 72
Ⓓ 8 x 8 = 64

13. Identify the multiplication sentence for the picture below:

Ⓐ 10 x 1 = 10
Ⓑ 9 x 2 = 18
Ⓒ 2 x 10 = 20
Ⓓ 5 x 4 = 20

14. Identify the multiplication sentence for the picture below:

Ⓐ 5 x 5 = 25
Ⓑ 4 x 4 = 16
Ⓒ 4 x 6 = 24
Ⓓ 5 x 4 = 20

15. Identify the multiplication sentence for the picture below

Ⓐ 3 x 1 = 3
Ⓑ 5 x 3 = 15
Ⓒ 3 x 2 = 6
Ⓓ 3 x 3 = 9

16. Represent the below equation as a multiplication expression. Write your answer in the box below.

8 + 8 + 8 + 8?

17. Match each multiplication statement to the correct addition statement by darkening the corresponding circles.

	Column A: 3+3+3+3+3+3	Column B: 3+3+3+3+3+3+3+3	Column C: 3+3+3
3 x 8	○	○	○
3 x 3	○	○	○
3 x 6	○	○	○

18. For each of the pictures, write the correct mathematical expression in the box.

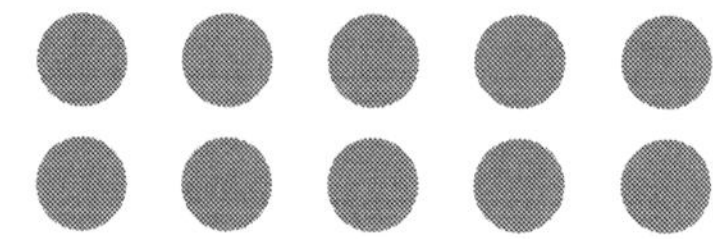

19. John finds the solution for 8 x 6 by solving for (8 x 5) + 8. Is John correct? Explain why you think that John's strategy is correct or not? Write your answer in the box below.

20. There are Seven boys, and each of them buys 6 pens. How many pens do they buy all together? Write an equation to represent this. Also, Find the total number of pens purchased using the equation.

21. Complete the following table:

Number of lions	5	6	9		
Total number of legs	20			32	16

Test ID	3M002
Test Name	Understanding Division
Student Name	
Date	

1	Ⓐ Ⓑ Ⓒ Ⓓ	6	Ⓐ Ⓑ Ⓒ Ⓓ	11	Ⓐ Ⓑ Ⓒ Ⓓ	16	Answer in the space provided below.
2	Ⓐ Ⓑ Ⓒ Ⓓ	7	Ⓐ Ⓑ Ⓒ Ⓓ	12	Ⓐ Ⓑ Ⓒ Ⓓ	17	Ⓐ Ⓑ Ⓒ Ⓓ
3	Ⓐ Ⓑ Ⓒ Ⓓ	8	Ⓐ Ⓑ Ⓒ Ⓓ	13	Ⓐ Ⓑ Ⓒ Ⓓ	18	Ⓐ Ⓑ Ⓒ
4	Ⓐ Ⓑ Ⓒ Ⓓ	9	Ⓐ Ⓑ Ⓒ Ⓓ	14	Ⓐ Ⓑ Ⓒ Ⓓ	19	Answer in the space provided below.
5	Ⓐ Ⓑ Ⓒ Ⓓ	10	Ⓐ Ⓑ Ⓒ Ⓓ	15	Ⓐ Ⓑ Ⓒ Ⓓ	20	Answer in the space provided below.

16

19

Dividend	Possible Divisors
100	2
100	4
100	
100	
100	20
100	25
100	

20

	<	>	=
30 ÷ 5 ____ 42 ÷ 6	○	○	○
72 ÷ 8 ____ 63 ÷ 7	○	○	○
54 ÷ 6 ____ 56 ÷ 7	○	○	○

Date of Completion:______________ Score:______________

Chapter 1 → Lesson 2: Understanding Division

1. George is canning pears. He has 100 pears and he divides the pears evenly among 10 pots. How many pears does George put in each pot?

   Ⓐ 9 pears
   Ⓑ 5 pears
   Ⓒ 8 pears
   Ⓓ 10 pears

2. Marisa made 15 woolen dolls. She gave the same number of woolen dolls to 3 friends. How many dolls did Marisa give to each friend?

   Ⓐ 4 woolen dolls
   Ⓑ 3 woolen dolls
   Ⓒ 5 woolen dolls
   Ⓓ 6 woolen dolls

3. Lisa bought 50 mangoes. She divided them equally into 5 basins. How many mangoes did Lisa put in each basin?

   Ⓐ 10 mangos
   Ⓑ 8 mangos
   Ⓒ 5 mangos
   Ⓓ 7 mangos

4. Jennifer picked 30 oranges from the basket. If it takes 6 oranges to make a one liter jar of juice, how many one liter jars of juice can Jennifer make?

   Ⓐ 4 jars
   Ⓑ 3 jars
   Ⓒ 6 jars
   Ⓓ 5 jars

5. Miller bought 80 rolls of paper towels. If there are 10 rolls of paper towels in each pack, how many packs of paper towels did Miller buy?

   Ⓐ 6 packs
   Ⓑ 8 packs
   Ⓒ 7 packs
   Ⓓ 5 packs

6. James takes 15 photographs of his school building. He gave the same number of photographs to 5 friends. How many photographs did James give to each friend?

   Ⓐ 2 photographs
   Ⓑ 3 photographs
   Ⓒ 6 photographs
   Ⓓ 5 photographs

7. Ron took 81 playing cards and arranged them into 9 equal piles. How many playing cards did Ron put in each pile?

   Ⓐ 5 playing cards
   Ⓑ 4 playing cards
   Ⓒ 6 playing cards
   Ⓓ 9 playing cards

8. Robert wants to buy 40 ice cream cups from the ice cream parlor. If there are 10 ice cream cups in each box, how many boxes of ice cream cups should Robert buy?

   Ⓐ 6 boxes
   Ⓑ 3 boxes
   Ⓒ 4 boxes
   Ⓓ 5 boxes

9. Marilyn wants to purchase 20 tiles. If the tiles come in packs of 5, how many packs should Marilyn buy?

   Ⓐ 3 packs
   Ⓑ 4 packs
   Ⓒ 5 packs
   Ⓓ 6 packs

10. There are 30 people running around the path. If the runners are evenly divided among the path's 5 lanes, how many people are running in each lane?

    Ⓐ 6 runners
    Ⓑ 5 runners
    Ⓒ 8 runners
    Ⓓ 4 runners

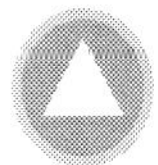

11. Sally is buying goodie bags for her class. She needs 24 bags in all. If the bags come in packs of 3, how many packs does Sally need?

Ⓐ 21 packs
Ⓑ 3 packs
Ⓒ 24 packs
Ⓓ 8 packs

12. Mr. Johnson is planting a garden. He wants to use all of his 44 seeds and wants to make 4 rows of vegetables. How many seeds should he plant in each row?

Ⓐ 22 seeds
Ⓑ 11 seeds
Ⓒ 4 seeds
Ⓓ 88 seeds

13. Destiny, Jimmy, and Marcy have 32 marbles all together. Tommy adds 4 marbles to the set. If the group of friends wants to evenly divide the marbles so that each person has the same number, how many marbles should each person receive?

Ⓐ 4 marbles
Ⓑ 8 marbles
Ⓒ 9 marbles
Ⓓ 10 marbles

14. Mr. Baker earned $100 for five days of work. If he made the same amount each day, how much money did he make per day?

Ⓐ $15 per day
Ⓑ $20 per day
Ⓒ $25 per day
Ⓓ $30 per day

15. Seth and his brother have collected 26 seashells on the beach. If they want to share them equally, how many seashells will each of them receive?

Ⓐ 6 seashells
Ⓑ 9 seashells
Ⓒ 26 seashells
Ⓓ 13 seashells

16. A pizza is cut into 8 slices. Tim and Kira want to share the pizza. If they both eat the same number of slices, how many slices will each person eat? Write it in the box given below.

17. Gabriela has 16 stickers. She wants to find two ways to divide the stickers into equal groups. Which expressions can she use to divide the stickers? Mark all the correct answers.

Ⓐ 16 ÷ 2
Ⓑ 16 ÷ 3
Ⓒ 16 ÷ 4
Ⓓ 16 ÷ 5

18. Circle the picture that shows the expression 10 ÷ 5.

Ⓐ 

Ⓑ 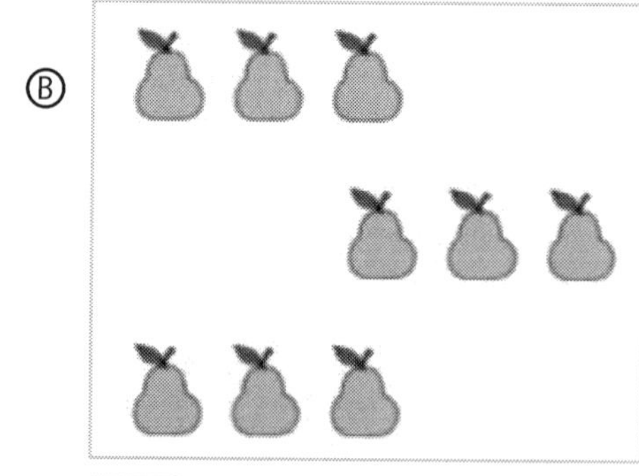

Ⓒ 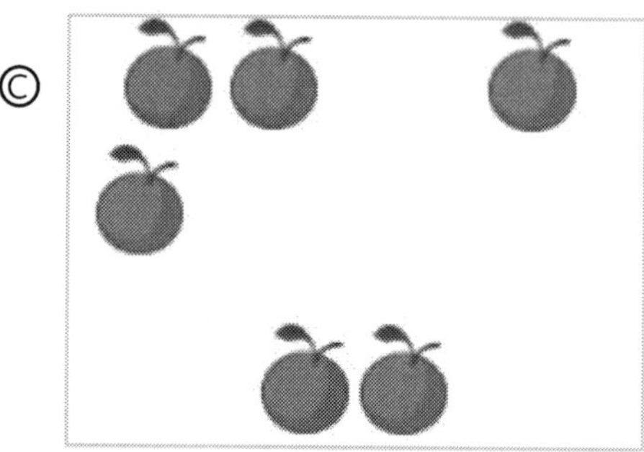

19. Miriam has 100 marbles. She wants to divide the marbles into equal groups. How many ways can she do this? Complete the table by listing all the possible ways in which you can divide 100. Write the missing numbers and fill the table. Enter the numbers in ascending order.

Dividend	Possible Divisors
100	2
100	4
100	
100	
100	20
100	25
100	

20. For each expression below, choose the correct symbol to be filled in the blank.

	<	>	=
30 ÷ 5 ____ 42 ÷ 6	○	○	○
72 ÷ 8 ____ 63 ÷ 7	○	○	○
54 ÷ 6 ____ 56 ÷ 7	○	○	○

Test ID	3M003
Test Name	Applying Multiplication & Division
Student Name	
Date	

1	Ⓐ Ⓑ Ⓒ Ⓓ	6	Ⓐ Ⓑ Ⓒ Ⓓ	11	Ⓐ Ⓑ Ⓒ Ⓓ	16	Ⓐ Ⓑ Ⓒ Ⓓ
2	Ⓐ Ⓑ Ⓒ Ⓓ	7	Ⓐ Ⓑ Ⓒ Ⓓ	12	Ⓐ Ⓑ Ⓒ Ⓓ	17	Ⓐ Ⓑ Ⓒ Ⓓ
3	Ⓐ Ⓑ Ⓒ Ⓓ	8	Ⓐ Ⓑ Ⓒ Ⓓ	13	Ⓐ Ⓑ Ⓒ Ⓓ	18	Answer in the space provided below.
4	Ⓐ Ⓑ Ⓒ Ⓓ	9	Ⓐ Ⓑ Ⓒ Ⓓ	14	Ⓐ Ⓑ Ⓒ Ⓓ	19	Answer in the space provided below.
5	Ⓐ Ⓑ Ⓒ Ⓓ	10	Ⓐ Ⓑ Ⓒ Ⓓ	15	Ⓐ Ⓑ Ⓒ Ⓓ	20	Answer in the space provided below.

18

19

20

Chapter 1 → Lesson 3: Applying Multiplication & Division

1. The product in this number sentence is ________.
   54 x 3 = ?

   Ⓐ 54
   Ⓑ 162
   Ⓒ 3
   Ⓓ 54 and 3

2. A Snack Shop has twice as many popcorn balls as they do cotton candy. If there are 30 popcorn balls, how many cotton candies are there?

   Ⓐ 7
   Ⓑ 450
   Ⓒ 30
   Ⓓ 15

3. Monica has 56 DVDs in her movie collection. This is 8 times as many as Sue has. How many DVDs does Sue have?

   Ⓐ 8
   Ⓑ 6
   Ⓒ 7
   Ⓓ 10

4. Jonathan can do 7 jumping jacks. Marcus can do 4 times as many as Jonathan. How many jumping jacks can Marcus do?

   Ⓐ 28
   Ⓑ 8
   Ⓒ 4
   Ⓓ 7

5. Darren has seen 4 movies this year. Marsha has seen 3 times as many movies as Darren. How many movies has Marsha seen?

   Ⓐ 7
   Ⓑ 3
   Ⓒ 4
   Ⓓ 12

6. **Sarah is planting a garden. She will plant 4 rows with 9 seeds in each row. How many plants will be in the garden?**

   Ⓐ 32 plants
   Ⓑ 36 plants
   Ⓒ 42 plants
   Ⓓ 13 plants

7. **Mrs. Huerta's class is having a pizza party. There are 24 students in the class. Each pizza has 12 slices. How many pizzas does Mrs. Huerta need to order for each child to have 1 slice?**

   Ⓐ 3 pizzas
   Ⓑ 2 pizzas
   Ⓒ 1 pizza
   Ⓓ 4 pizzas

8. **There are 27 apples. How many pies can be made if each pie uses 3 apples?**

   Ⓐ 7 pies
   Ⓑ 8 pies
   Ⓒ 9 pies
   Ⓓ 10 pies

9. **Keegan is planting a garden in even rows. He has 48 seeds. Which layout is NOT possible?**

   Ⓐ 6 rows of 8 seeds
   Ⓑ 8 rows of 6 seeds
   Ⓒ 7 rows of 7 seeds
   Ⓓ 12 rows of 4 seeds

10. **There are 25 students in a gym class. They want to play a game with 5 equal teams. How many students will be on each team?**

    Ⓐ 4 students
    Ⓑ 5 students
    Ⓒ 7 students
    Ⓓ 3 students

11. Josie has 7 days to read a book with 21 chapters. How many chapters should she read each day?

Ⓐ 3 chapters
Ⓑ 4 chapters
Ⓒ 5 chapters
Ⓓ 7 chapters

12. Devon has $40 to spend on fuel. One gallon of fuel costs $5. How many gallons can Devon afford to buy?

Ⓐ 5 gallons
Ⓑ 12 gallons
Ⓒ 9 gallons
Ⓓ 8 gallons

13. Amanda is using the following cake recipe:
4 cups flour
1 cup sugar
3 cups milk
1 egg
If Amanda needs to make three batches, how many cups of flour will she need?

Ⓐ 7 cups
Ⓑ 12 cups
Ⓒ 10 cups
Ⓓ 16 cups

14. Kim invited 20 friends to her birthday party. Twice as many friends than she invited showed up the day of the party. Which number sentence could be used to solve how many friends came to the party?

Ⓐ n + 20 = 2
Ⓑ n x 2 = 20
Ⓒ 20 x 2 = n
Ⓓ 20 - n = 20

15. The product of 9 and a number is 45.
Which number sentence models this situation?

Ⓐ 9 + n = 45
Ⓑ 45 + 9 = n
Ⓒ 9 x n = 45
Ⓓ 5 x n = 45

16. A classroom has 5 rows of desks. There are 6 desks in each row. How many desks are there altogether? Select the number sentences that represent the solution. Choose all correct answers.

Ⓐ 6 - 5= 1
Ⓑ 5 x 6= 30
Ⓒ 6 x 5= 30
Ⓓ 5 + 6= 11

17. Jasmine bought a bouquet of 24 flowers. She plans to give the same number of flowers to her 4 friends, Daniel, Raquel, Elliot and Sue. How many flowers will each friend receive? Circle the correct answer.

Ⓐ 2
Ⓑ 5
Ⓒ 6
Ⓓ 4

18. There are 48 cupcakes to be shared equally among 6 boys. How many cupcakes will each boy get? Write your answer in the box given below.

19. Fill in the blank with the correct operational symbol to make these equations true.

A) 64 ÷ 8 = 2 ___ 4.

B) 2 ___ 4 = 42 ÷ 7.

20. Joseph reads 8 pages every day. In how many days will he be able to complete reading a book which has 56 pages? Write an equation to represent this in the box below, and find the number of days Joseph takes to complete the book.

Test ID	3M004
Test Name	Finding Unknown Values
Student Name	
Date	

1	(A) (B) (C) (D)	6	(A) (B) (C) (D)	11	(A) (B) (C) (D)	16	Answer in the space provided below.
2	(A) (B) (C) (D)	7	(A) (B) (C) (D)	12	(A) (B) (C) (D)	17	(A) (B) (C) (D)
3	(A) (B) (C) (D)	8	(A) (B) (C) (D)	13	(A) (B) (C) (D)	18	Answer in the space provided below.
4	(A) (B) (C) (D)	9	(A) (B) (C) (D)	14	(A) (B) (C) (D)	19	Answer in the space provided below.
5	(A) (B) (C) (D)	10	(A) (B) (C) (D)	15	(A) (B) (C) (D)	20	Answer in the space provided below.

16

18

Equation	Product
4 x 8=	
	x 7 = 63
3 x 5=	
	x 1 =7

19

20

Equation	n=7	n=6
$3 \times 8 = 4 \times n$		
$72 \div 9 = 56 \div n$		

Date of Completion:______________ Score:______________

Chapter 1 → Lesson 4: Finding Unknown Values

1. Find the number that makes this equation true.
   n x 6 = 30

   Ⓐ n = 11
   Ⓑ n = 7
   Ⓒ n = 5
   Ⓓ n = 3

2. Find the number that makes this equation true.
   7 x ___ = 21

   Ⓐ 3
   Ⓑ 4
   Ⓒ 5
   Ⓓ 6

3. Find the number that makes this equation true.
   ___ x 4 = 36

   Ⓐ 9
   Ⓑ 8
   Ⓒ 7
   Ⓓ 6

4. Find the number that makes this equation true.
   n ÷ 9 = 8

   Ⓐ n = 81
   Ⓑ n = 45
   Ⓒ n = 72
   Ⓓ n = 63

5. Find the number that makes this equation true.
   ____ ÷ 3 = 10

   Ⓐ 27
   Ⓑ 30
   Ⓒ 33
   Ⓓ 60

6. Find the number that makes this equation true.
   45 ÷ n = 9

   Ⓐ n = 10
   Ⓑ n = 7
   Ⓒ n = 5
   Ⓓ n = 3

7. Find the number that makes this equation true.
   64 = ___ x 8

   Ⓐ 6
   Ⓑ 7
   Ⓒ 8
   Ⓓ 9

8. Find the number that makes this equation true.
   12 ÷ ___ = 2

   Ⓐ 7
   Ⓑ 6
   Ⓒ 8
   Ⓓ 4

9. Find the number that makes this equation true.
   ___ ÷ 7 = 11

   Ⓐ 63
   Ⓑ 70
   Ⓒ 77
   Ⓓ 78

10. Find the number that makes this equation true.
    16 = n x 4

    Ⓐ n = 12
    Ⓑ n = 4
    Ⓒ n = 3
    Ⓓ n = 2

11. For what value of m is this equation true?
m x 7 = 56

Ⓐ m = 6
Ⓑ m = 8
Ⓒ m = 7
Ⓓ m = 12

12. For what value of n is this equation true?
60 ÷ n = 5

Ⓐ n = 8
Ⓑ n = 12
Ⓒ n = 14
Ⓓ n = 16

13. For what value of z is this equation true?
9 x z = 81

Ⓐ z = 11
Ⓑ z = 9
Ⓒ z = 12
Ⓓ z = 7

14. For what value of p is this equation true?
p ÷ 3 = 3

Ⓐ p = 6
Ⓑ p = 1
Ⓒ p = 9
Ⓓ p = 0

15. For what value of u is this equation true?
10 ÷ u = 10

Ⓐ u = 1
Ⓑ u = 0
Ⓒ u = 10
Ⓓ u = 100

16. Which number makes the equation true? Write your answer in the box given.

____ ÷ 6 = 9

17. Which two numbers will make the equation true?

___ x ___ = 14

Ⓐ 5
Ⓑ 7
Ⓒ 6
Ⓓ 2

18. Enter the correct answer in the table.

Equation	Product
4 x 8=	
	x 7 = 63
3 x 5=	
	x 1 =7

19. A pen costs $7 to buy. How much would six pens cost? Write an equation to represent this in the box below.

20. Match the value of n for each of the equations given.

Equation	n=7	n=6
$3 \times 8 = 4 \times n$	○	○
$72 \div 9 = 56 \div n$	○	○

Test ID	3M005
Test Name	Multiplication & Division Properties
Student Name	
Date	

1	Ⓐ Ⓑ Ⓒ Ⓓ	6	Ⓐ Ⓑ Ⓒ Ⓓ	11	Ⓐ Ⓑ Ⓒ Ⓓ	16	Ⓐ Ⓑ Ⓒ Ⓓ
2	Ⓐ Ⓑ Ⓒ Ⓓ	7	Ⓐ Ⓑ Ⓒ Ⓓ	12	Ⓐ Ⓑ Ⓒ Ⓓ	17	Answer in the space provided below.
3	Ⓐ Ⓑ Ⓒ Ⓓ	8	Ⓐ Ⓑ Ⓒ Ⓓ	13	Ⓐ Ⓑ Ⓒ Ⓓ	18	Ⓐ Ⓑ Ⓒ Ⓓ
4	Ⓐ Ⓑ Ⓒ Ⓓ	9	Ⓐ Ⓑ Ⓒ Ⓓ	14	Ⓐ Ⓑ Ⓒ Ⓓ	19	Answer in the space provided below.
5	Ⓐ Ⓑ Ⓒ Ⓓ	10	Ⓐ Ⓑ Ⓒ Ⓓ	15	Ⓐ Ⓑ Ⓒ Ⓓ	20	Answer in the space provided below.

17

19

20

	3 x (5 x 7) = (3 x 5) x 7	3 x 1 = 3	3 x 5 = 5 x 3	3 x (5 + 7) = (3 x 5) + (3 x 7)
Commutative Property	○	○	○	○
Associative Property	○	○	○	○
Identity Property	○	○	○	○
Distributive Property	○	○	○	○

Date of Completion:______________ Score:______________

Chapter 1 → Lesson 5: Multiplication & Division Properties

1. Which of these statements is not true?

   Ⓐ 4 x (3 x 6) = (4 x 3) x 6
   Ⓑ 4 x 3 = 3 x 4
   Ⓒ 15 x 0 = 0 x 15
   Ⓓ 12 x 1 = 12 x 12

2. Which of these statements is true?

   Ⓐ The product of 11 x 6 is equal to the product of 6 x 11.
   Ⓑ The product of 11 x 6 is greater than the product of 6 x 11.
   Ⓒ The product of 11 x 6 is less than the product of 6 x 11.
   Ⓓ There is no relationship between the product of 11 x 6 and the product of 6 x 11.

3. Which of the following expressions has a value of 0?

   Ⓐ (3 x 4) x 1
   Ⓑ 50 x 1
   Ⓒ 3 x 4 x 0
   Ⓓ (3 x 1) x 2

4. Select the option in which both the numerical expressions result in a value of 0?

   Ⓐ 60 x 1 and 1 x 60
   Ⓑ 10 x 10 and 0 x 10
   Ⓒ 27 x 0 and 0 x 27
   Ⓓ 0 ÷ 15 and 15 ÷ 15

5. Which mathematical property does this equation model?
   6 x 1 = 6

   Ⓐ Commutative Property of Multiplication
   Ⓑ Associative Property of Multiplication
   Ⓒ Identity Property of Multiplication
   Ⓓ Distributive Property

6. Which mathematical property does this equation model?
9 x 6 = 6 x 9

Ⓐ Commutative Property of Multiplication
Ⓑ Associative Property of Multiplication
Ⓒ Identity Property of Multiplication
Ⓓ Distributive Property

7. Which mathematical property does this equation model?
(2 x 10) x 3 = 2 x (10 x 3)

Ⓐ Commutative Property of Multiplication
Ⓑ Associative Property of Multiplication
Ⓒ Identity Property of Multiplication
Ⓓ Distributive Property

8. Which mathematical property does this equation model?
4 x (9 + 6) = (4 x 9) + (4 x 6)

Ⓐ Commutative Property of Multiplication
Ⓑ Associative Property of Multiplication
Ⓒ Identity Property of Multiplication
Ⓓ Distributive Property

9. By the Commutative Property of Multiplication, if you know that 4 x 5= 20, then you also know that ______ .

Ⓐ 20 is an even number
Ⓑ 4 x 6 = 24
Ⓒ 5 x 4 = 20
Ⓓ 5 is greater than 4

10. By the Associative Property of Multiplication, If you know that (2 x 3) x 4 = 24, then you also know that __________.

Ⓐ 2 x (3 x 4) = 24
Ⓑ 2 x 4 = 8
Ⓒ 24 ÷ 6 = 4
Ⓓ (2 x 3) x 5 = 30

11. Complete the following statement:
Multiplication and _______ are inverse operations.

Ⓐ addition
Ⓑ subtraction
Ⓒ division
Ⓓ distribution

12. Below equation models the ___________.
32 x 7 = 7 x 32

Ⓐ Commutative Property of Multiplication
Ⓑ Associative Property of Multiplication
Ⓒ Identity Property of Multiplication
Ⓓ Distributive Property

13. Below equation models the ___________ .
26 x 2 = (20 x 2) + (6 x 2)

Ⓐ Commutative Property of Multiplication
Ⓑ Associative Property of Multiplication
Ⓒ Identity Property of Multiplication
Ⓓ Distributive Property

14. By the Identity Property of Multiplication, you know that __________.

Ⓐ 2 x 2 = 4
Ⓑ 0 x 0 = 0
Ⓒ 6 x 1 = 6
Ⓓ 5 ÷ 5 = 1

15. What number belongs in the blank?
10 x __ = 10

Ⓐ 1
Ⓑ 0
Ⓒ 10
Ⓓ 5

16. From the below 4 options, select 2 options that will result in the same product based on the commutative property of multiplication?

Ⓐ 5 + 4
Ⓑ 10 x 2
Ⓒ 5 x 4
Ⓓ 4 x 5

17. Make the equation true according to the Identity Property of Multiplication. Write the correct number in the answer box given below.

7 x ___ = 7

18. (2 x 3) x 4 = 24 and 2 x (3 x 4) = 24

Identify the property that is applicable. Circle the correct answer choice.

Ⓐ Associative property of multiplication
Ⓑ Distributive property
Ⓒ Commutative property of multiplication
Ⓓ Identity Property of Multiplication

19. Which number will make the below equation true. Write your answer in the box given below.

3 x 7 = 7 x ?

20. Match the property with the correct example by shading the appropriate circle.

	3 x (5 x 7) = (3 x 5) x 7	3 x 1 = 3	3 x 5 = 5 x 3	3 x (5 + 7) = (3 x 5) + (3 x 7)
Commutative Property	◯	◯	◯	◯
Associative Property	◯	◯	◯	◯
Identity Property	◯	◯	◯	◯
Distributive Property	◯	◯	◯	◯

Test ID	3M006
Test Name	Relating Multiplication & Division
Student Name	
Date	

1	Ⓐ Ⓑ Ⓒ Ⓓ	6	Ⓐ Ⓑ Ⓒ Ⓓ	11	Ⓐ Ⓑ Ⓒ Ⓓ	16	Answer in the space provided below.
2	Ⓐ Ⓑ Ⓒ Ⓓ	7	Ⓐ Ⓑ Ⓒ Ⓓ	12	Ⓐ Ⓑ Ⓒ Ⓓ	17	Answer in the space provided below.
3	Ⓐ Ⓑ Ⓒ Ⓓ	8	Ⓐ Ⓑ Ⓒ Ⓓ	13	Ⓐ Ⓑ Ⓒ Ⓓ	18	Ⓐ Ⓑ Ⓒ Ⓓ
4	Ⓐ Ⓑ Ⓒ Ⓓ	9	Ⓐ Ⓑ Ⓒ Ⓓ	14	Ⓐ Ⓑ Ⓒ Ⓓ	19	Ⓐ Ⓑ Ⓒ Ⓓ
5	Ⓐ Ⓑ Ⓒ Ⓓ	10	Ⓐ Ⓑ Ⓒ Ⓓ	15	Ⓐ Ⓑ Ⓒ Ⓓ	20	Answer in the space provided below.

16

	40÷5=8	36÷6=6	21÷7=3	18÷9=2
6 x 6= 36				
9 x 2= 18				
5 x 8= 40				
7 x 3= 21				

17

	Division Expression	Quotient
If 4 x 4= 16 then...	16 ÷ 4=	
If 7 x 6= 42 then... 42 ÷		6
If 3 x 9= 27 then...		÷ 3=9

20

Date of Completion:_____________ Score:_____________

Chapter 1 → Lesson 6: Relating Multiplication & Division

1. Find the number that would complete both of the following number sentences.
   ___ x 6 = 30
   30 ÷ 6 = ___

   Ⓐ 7
   Ⓑ 5
   Ⓒ 6
   Ⓓ 24

2. Find the number that would complete both of the following number sentences.
   7 x ___ = 21
   21 ÷ ___ = 7

   Ⓐ 5
   Ⓑ 14
   Ⓒ 3
   Ⓓ 7

3. Find the number that would complete both of the following number sentences.
   72 ÷ ___ = 8
   8 x ___ = 72

   Ⓐ 8
   Ⓑ 9
   Ⓒ 10
   Ⓓ 64

4. Find the number that would complete both of the following number sentences.
   50 ÷ ___ = 5
   ___ x 5 = 50

   Ⓐ 15
   Ⓑ 5
   Ⓒ 45
   Ⓓ 10

5. Find the number that would complete both of the following number sentences.
36 ÷ ____ =
_____ x 9 = 36

Ⓐ 4
Ⓑ 5
Ⓒ 27
Ⓓ 9

6. There are 9 students in a group. Each student needs 5 sheets of paper to complete a project. Which number sentence below can be used to find out how many total sheets of paper are needed for this project? Select all the correct answer choices.

Ⓐ 5 x ____ = 9
Ⓑ 9 x 5 = ____
Ⓒ ____ ÷ 5 = 9
Ⓓ 9 ÷ 5 = ____

7. In a football game, Timmy scored 8 touchdowns. Each touchdown was worth 7 points. Which number sentence below can be used to find out how many points Timmy scored in all?

Ⓐ 56 x ___ = 8
Ⓑ ____ ÷ 7 = 8
Ⓒ 7 x 56 = ____
Ⓓ 8 ÷ 7 = _____

8. Devon needs to buy 96 pencils for his goodie bags. Pencils are sold in packages of 12. Which number sentence below can be used to find out how many packages Devon needs to buy?

Ⓐ 12 ÷ 96 = ___
Ⓑ 12 x 96 = ____
Ⓒ ____ x 12 = 96
Ⓓ 12 ÷ ____ = 96

9. Eighty-four students are attending an awards ceremony. They are to be seated at twelve equal tables. Which number sentence below can be used to find out how many students should be assigned to each table?

Ⓐ 6 x ____ = 84
Ⓑ 12 x ____ = 84
Ⓒ 84 x 12 = _____
Ⓓ 12 ÷ 84 = ____

10. Walter has 16 slices of pizza to share among himself and seven friends. He wants each person to get an equal number of slices. Which number sentence below can be used to find out how many slices each person will get?

Ⓐ 7 ÷ 16 = ____
Ⓑ 7 x 16 = ____
Ⓒ 7 x ____ = 16
Ⓓ 8 x ____ = 16

11. Which of the following equations would be in the same fact family as:
6 x 5 = 30?

Ⓐ 30 ÷ 10 = 3
Ⓑ 6 x 30 = 5
Ⓒ 30 ÷ 5 = 6
Ⓓ 5 ÷ 30 = 6

12. Which number sentence is equivalent to the number sentence below?
4 x n = 32

Ⓐ 4 x 32 = n
Ⓑ n + 4 = 32
Ⓒ 32 ÷ 4 = n
Ⓓ 4 x 4 = n

13. Which number sentence is equivalent to the number sentence below?
n x 6 = 48

Ⓐ 48 x n = 6
Ⓑ 6 x 4 = n
Ⓒ 48 x 6 = n
Ⓓ 48 ÷ n = 6

14. Which number sentence is equivalent to the number sentence below?
45 ÷ n = 9

Ⓐ 9 x 45 = n
Ⓑ 45 x n = 9
Ⓒ 9 x n = 45
Ⓓ n + 9 = 45

15. David receives 2 pieces of candy for each chore that he completes each week. This week he earned 32 pieces of candy. Which number sentence below can be used to figure out how many chores David completed?

Ⓐ 2 x 32 = ___
Ⓑ ___ x 2 = 32
Ⓒ 32 + 2 = ___
Ⓓ 2 ÷ 32 = ___

16. Match each multiplication sentence with the corresponding division sentence. Remember, if a x b= c then c ÷ a= b.

	40÷5=8	36÷6=6	21÷7=3	18÷9=2
6 x 6= 36	○	○	○	○
9 x 2= 18	○	○	○	○
5 x 8= 40	○	○	○	○
7 x 3= 21	○	○	○	○

17. Complete the table by writing the correct number. Remember, if a x b= c then c ÷ a= b.

	Division Expression	Quotient
If 4 x 4= 16 then...	16 ÷ 4=	
If 7 x 6= 42 then... 42 ÷		6
If 3 x 9= 27 then...		÷ 3=9

18. Which multiplication sentences relate to 63 ÷ 9= 7? Select all the correct answers. Note: More than one option may be correct.

Ⓐ 7 x 7= 49
Ⓑ 7 x 9= 63
Ⓒ 9 x 7= 63
Ⓓ 9 x 8= 72

19. Which number sentence is equivalent to the number sentence below:

$65 \div n = 5$.

Ⓐ $65 = 5 \times n$
Ⓑ $65 = 5 \div n$
Ⓒ $65 = n \div 5$
Ⓓ $65 = 5 - n$

20. Fill in the blanks with "<, > or =" to complete the following expressions.

A) $40 \div 5$ ____ $54 \div 9$

B) $35 \div 7$ ____ $28 \div 4$

C) $18 \div 6$ ____ $24 \div 8$

Test ID	3M007
Test Name	Multiplication & Division Facts
Student Name	
Date	

1	Ⓐ Ⓑ Ⓒ Ⓓ	11	Ⓐ Ⓑ Ⓒ Ⓓ	21	Ⓐ Ⓑ Ⓒ Ⓓ	31	Answer in the space provided below.
2	Ⓐ Ⓑ Ⓒ Ⓓ	12	Ⓐ Ⓑ Ⓒ Ⓓ	22	Ⓐ Ⓑ Ⓒ Ⓓ	32	Answer in the space provided below.
3	Ⓐ Ⓑ Ⓒ Ⓓ	13	Ⓐ Ⓑ Ⓒ Ⓓ	23	Ⓐ Ⓑ Ⓒ Ⓓ	33	Ⓐ Ⓑ Ⓒ Ⓓ
4	Ⓐ Ⓑ Ⓒ Ⓓ	14	Ⓐ Ⓑ Ⓒ Ⓓ	24	Ⓐ Ⓑ Ⓒ Ⓓ	34	Ⓐ Ⓑ Ⓒ Ⓓ
5	Ⓐ Ⓑ Ⓒ Ⓓ	15	Ⓐ Ⓑ Ⓒ Ⓓ	25	Ⓐ Ⓑ Ⓒ Ⓓ	35	Answer in the space provided below.
6	Ⓐ Ⓑ Ⓒ Ⓓ	16	Ⓐ Ⓑ Ⓒ Ⓓ	26	Ⓐ Ⓑ Ⓒ Ⓓ		
7	Ⓐ Ⓑ Ⓒ Ⓓ	17	Ⓐ Ⓑ Ⓒ Ⓓ	27	Ⓐ Ⓑ Ⓒ Ⓓ		
8	Ⓐ Ⓑ Ⓒ Ⓓ	18	Ⓐ Ⓑ Ⓒ Ⓓ	28	Ⓐ Ⓑ Ⓒ Ⓓ		
9	Ⓐ Ⓑ Ⓒ Ⓓ	19	Ⓐ Ⓑ Ⓒ Ⓓ	29	Ⓐ Ⓑ Ⓒ Ⓓ		
10	Ⓐ Ⓑ Ⓒ Ⓓ	20	Ⓐ Ⓑ Ⓒ Ⓓ	30	Ⓐ Ⓑ Ⓒ Ⓓ		

31

	12	18	32
6 x 3=	○	○	○
8 x 4=	○	○	○
4 x 3=	○	○	○
9 x 2=	○	○	○

32

35

5	x	8	=	
8	÷		=	8
	÷	7	=	0
6	x		=	30

Date of Completion:_______________ Score:_______________

Chapter 1 → Lesson 7: Multiplication & Division Facts

1. Find the product.
   6 x 0 = ____

   Ⓐ 6
   Ⓑ 1
   Ⓒ 0
   Ⓓ 2

2. Find the product.
   1 x 10 = ____

   Ⓐ 0
   Ⓑ 1
   Ⓒ 10
   Ⓓ 11

3. Solve.
   3 x 8 = ____

   Ⓐ 24
   Ⓑ 21
   Ⓒ 18
   Ⓓ 28

4. Solve.
   ___ = 5 x 9

   Ⓐ 40
   Ⓑ 45
   Ⓒ 50
   Ⓓ 35

5. Find the product of 8 and 6.

   Ⓐ 14
   Ⓑ 42
   Ⓒ 48
   Ⓓ 56

6. Find the product of 7 and 7.

   Ⓐ 42
   Ⓑ 46
   Ⓒ 49
   Ⓓ 56

7. Find the product of 4 and 6.

   Ⓐ 20
   Ⓑ 24
   Ⓒ 28
   Ⓓ 32

8. Find the product.
   6 x 9 = ____

   Ⓐ 54
   Ⓑ 45
   Ⓒ 48
   Ⓓ 64

9. Find the product.
   ___ = 9 x 8

   Ⓐ 64
   Ⓑ 72
   Ⓒ 81
   Ⓓ 82

10. Which expression below has a product of 48?

   Ⓐ 6 x 7
   Ⓑ 4 x 14
   Ⓒ 7 x 8
   Ⓓ 8 x 6

11. Find the quotient of 25 and 5.

   Ⓐ 20
   Ⓑ 5
   Ⓒ 4
   Ⓓ 15

12. What is 32 divided by 4?

Ⓐ 9
Ⓑ 8
Ⓒ 7
Ⓓ 6

13. What is 28 divided by 7?

Ⓐ 4
Ⓑ 5
Ⓒ 3
Ⓓ 6

14. Find the quotient.
$0 \div 5 =$ ____

Ⓐ 0
Ⓑ 1
Ⓒ 5
Ⓓ 50

15. Find the quotient.
$7 \div 1 =$ ____

Ⓐ 1
Ⓑ 0
Ⓒ 7
Ⓓ 8

16. Find the quotient.
_____ $= 12 \div 2$

Ⓐ 9
Ⓑ 8
Ⓒ 7
Ⓓ 6

17. Divide.
$63 \div 9 =$ ____

Ⓐ 6
Ⓑ 7
Ⓒ 8
Ⓓ 9

18. Divide.
42 ÷ 7 = ____

Ⓐ 5
Ⓑ 6
Ⓒ 7
Ⓓ 8

19. Find the quotient of 33 and 3.

Ⓐ 11
Ⓑ 12
Ⓒ 10
Ⓓ 9

20. Divide.
56 ÷ 7 = ____

Ⓐ 6
Ⓑ 7
Ⓒ 8
Ⓓ 9

21. Solve.
4 x 12 = _____

Ⓐ 36
Ⓑ 48
Ⓒ 42
Ⓓ 46

22. Solve.
____ = 75 ÷ 5

Ⓐ 13
Ⓑ 15
Ⓒ 17
Ⓓ 25

23. Solve.
84 ÷ 12 = _____

Ⓐ 7
Ⓑ 8
Ⓒ 9
Ⓓ 12

24. Solve.
12 x 3 = ____

Ⓐ **32**
Ⓑ **36**
Ⓒ **39**
Ⓓ **48**

25. Solve.
36 ÷ 3 = _____

Ⓐ **22**
Ⓑ **12**
Ⓒ **14**
Ⓓ **18**

26. Solve.
60 ÷ 5 = ____

Ⓐ **8**
Ⓑ **9**
Ⓒ **12**
Ⓓ **14**

27. Solve.
11 x 4 = ____

Ⓐ **32**
Ⓑ **44**
Ⓒ **39**
Ⓓ **46**

28. Solve.
_____ = 80 ÷ 8

Ⓐ **10**
Ⓑ **9**
Ⓒ **8**
Ⓓ **12**

29. Solve.
    12 x 8 = ____

    Ⓐ 72
    Ⓑ 84
    Ⓒ 92
    Ⓓ 96

30. Solve.
    50 ÷ 5 = ____

    Ⓐ 25
    Ⓑ 20
    Ⓒ 10
    Ⓓ 50

31. Match each equation to the correct product by shading the appropriate circle.

	12	18	32
6 x 3=	○	○	○
8 x 4=	○	○	○
4 x 3=	○	○	○
9 x 2=	○	○	○

32. Solve and write the answer in the box given below.

    24 ÷ 4=?

33. Which expressions have a product of 36. Select all the correct answers.

    Ⓐ 6 x 6
    Ⓑ 8 x 4
    Ⓒ 6 x 8
    Ⓓ 9 x 4

34. Find the quotient.

27 ÷ 9

Ⓐ 18
Ⓑ 4
Ⓒ 3
Ⓓ 2

35. Complete the following table.

5	x	8	=	
8	÷		=	8
	÷	7	=	0
6	x		=	30

Test ID	3M008
Test Name	Two-Step Problems
Student Name	
Date	

1	Ⓐ Ⓑ Ⓒ Ⓓ	6	Ⓐ Ⓑ Ⓒ Ⓓ	11	Ⓐ Ⓑ Ⓒ Ⓓ	16	Ⓐ Ⓑ Ⓒ Ⓓ
2	Ⓐ Ⓑ Ⓒ Ⓓ	7	Ⓐ Ⓑ Ⓒ Ⓓ	12	Ⓐ Ⓑ Ⓒ Ⓓ	17	Answer in the space provided below.
3	Ⓐ Ⓑ Ⓒ Ⓓ	8	Ⓐ Ⓑ Ⓒ Ⓓ	13	Ⓐ Ⓑ Ⓒ Ⓓ	18	Ⓐ Ⓑ Ⓒ Ⓓ
4	Ⓐ Ⓑ Ⓒ Ⓓ	9	Ⓐ Ⓑ Ⓒ Ⓓ	14	Ⓐ Ⓑ Ⓒ Ⓓ	19	Answer in the space provided below.
5	Ⓐ Ⓑ Ⓒ Ⓓ	10	Ⓐ Ⓑ Ⓒ Ⓓ	15	Ⓐ Ⓑ Ⓒ Ⓓ		

17

19

	$50	$5	$10	$4
Karen had 86 dollars. He bought 7 books. After buying them he had 16 dollars. How much did each book cost ?	○	○	○	○
Jose and his four friends bought a new board game. It was on sale for 20 dollars off. If each of the boys (total 5 of them) paid $6. What was the original cost of the new board game?	○	○	○	○
A shopkeeper buys 5 pens for $35 and sells them at the rate of $8 per pen. If he sells all the five pens, how much profit he will get?	○	○	○	○
Jeffrey bought 8 actions figures which cost 3 dollars each from John. John bought 6 books from the amount he received from Jeffrey. If the cost of each book John purchased is the same, what is the cost of each book?	○	○	○	○

Chapter 1 → Lesson 8: Two-Step Problems

1. Danny has 47 baseball cards. He gives his brother 11 cards. Danny then divides the remaining cards between 3 of his classmates. How many cards does each classmate receive?

   Ⓐ 15
   Ⓑ 3
   Ⓒ 12
   Ⓓ 11

2. Two classes of grade three students are lined up outside. One class is lined up in 3 rows of 7. The other class is lined up in 4 rows of 5. How many total third graders are lined up outside?

   Ⓐ 19 third graders
   Ⓑ 21 third graders
   Ⓒ 41 third graders
   Ⓓ 20 third graders

3. Jessica earns 10 dollars per hour for babysitting. She has saved 60 dollars so far. How many more hours will she need to babysit to buy something that costs 100 dollars?

   Ⓐ 40 hours
   Ⓑ 6 hours
   Ⓒ 10 hours
   Ⓓ 4 hours

4. George started with 2 bags of 10 cookies. He gave 12 cookies to his parents. How many cookies does George have now?

   Ⓐ 8 cookies
   Ⓑ 10 cookies
   Ⓒ 12 cookies
   Ⓓ 20 cookies

5. Renae has 60 minutes to do her chores and do her homework. She has 3 chores to complete and each chore takes 15 minutes to complete. After completing her chores, how many minutes does Renae have left to do her homework?

   Ⓐ 15 minutes
   Ⓑ 45 minutes
   Ⓒ 30 minutes
   Ⓓ 0 minutes

6. Anna and Jamie want to buy a new board game. The original cost was 28 dollars. It is on sale for 4 dollars off. How much money should each girl pay if they buy the game on sale and pay equal amounts?

   Ⓐ $24
   Ⓑ $2
   Ⓒ $12
   Ⓓ $14

7. 100 students went on a field trip. Ten students rode with their parents in a car while the remaining students were divided equally into 5 buses. How many students rode on each bus?

   Ⓐ 9 students
   Ⓑ 18 students
   Ⓒ 50 students
   Ⓓ 90 students

8. Julia has 32 books. Her sister has twice the number of books that Julia has. How many books do the girls have altogether?

   Ⓐ 66 books
   Ⓑ 32 books
   Ⓒ 64 books
   Ⓓ 96 books

9. Alicia bought 5 crates of apples. Each crate had 8 apples. She divided the apples equally into 10 bags. How many apples were in each bag?

   Ⓐ 40 apples
   Ⓑ 4 apples
   Ⓒ 10 apples
   Ⓓ 5 apples

10. Janeth went to the store and spent 4 dollars on markers. She also bought 3 copies of the same book. If she spent a total of 19 dollars, how much did each book cost?

    Ⓐ 5 dollars
    Ⓑ 4 dollars
    Ⓒ 3 dollars
    Ⓓ 6 dollars

11. Brian won 24 candy bars in a contest. He gave 2 candy bars to each of his 7 friends. How many candy bars does Brian have left?

Ⓐ 14 candy bars
Ⓑ 12 candy bars
Ⓒ 10 candy bars
Ⓓ 17 candy bars

12. Jenine gave 3 mini cupcakes to each of her three sisters. She then had 4 left for herself. How many mini cupcakes did Jenine start with?

Ⓐ 13 mini cupcakes
Ⓑ 9 mini cupcakes
Ⓒ 10 mini cupcakes
Ⓓ 7 mini cupcakes

13. Twenty-two people visited the art exhibit at the museum on Friday. Twice as many people visited on Saturday. How many people combined visited the art exhibit at the museum on Friday and Saturday?

Ⓐ 88 people
Ⓑ 66 people
Ⓒ 22 people
Ⓓ 44 people

14. Audrey can watch 5 hours of TV a week. She has already watched 4 shows that are each 1 hour long. How many more hours can she watch TV this week?

Ⓐ 3 hours
Ⓑ 2 hours
Ⓒ 1 hour
Ⓓ 4 hours

15. Greg had 3 books. His older brother gave him 15 more books. Greg wants to divide his total number of books equally onto 6 shelves. How many books should he place on each shelf?

Ⓐ 3 books
Ⓑ 12 books
Ⓒ 18 books
Ⓓ 6 books

16. Sarah bought 2 boxes of doughnuts. Each box contained 12 donuts. She shared a total of 7 donuts with her friends. How many doughnuts does she have now? Identify which equations can be used to find the answer. (Choose all correct answers)

Ⓐ 12 x 2= 24
Ⓑ 12 + 2= 14
Ⓒ 24-7= 17
Ⓓ 2 + 12 + 7= 21

17. Freddy has a collection of 32 baseball cards. He wants to share the cards with 4 classmates. One of the classmates brings 8 more cards to add to the collection. If Freddy and his classmates share all the cards, each receiving the same number, how many cards does each person have? What should be the steps to be followed to arrive at the answer? Write the steps in the correct sequence in the boxes given below.

Ⓐ 32 - 6 = 26
Ⓑ 4 + 1 = 5
Ⓒ 32 + 8 = 40
Ⓓ 40 ÷ 5 = 8

18. A farmer collected 22 pints of milk from his cows. He put all the milk into bottles. Each bottle holds 2 pints of milk. He accidentally spilled 6 bottles of milk. How many bottles are left with the farmer now? Circle the math sentences that can be used to find the answer. (Circle all correct answers)

Ⓐ 11
Ⓑ 24
Ⓒ 16
Ⓓ 5

19. For each of the problems in the first column, select the correct answer.

	$50	$5	$10	$4
Karen had 86 dollars. He bought 7 books. After buying them he had 16 dollars. How much did each book cost ?	◯	◯	◯	◯
Jose and his four friends bought a new board game. It was on sale for 20 dollars off. If each of the boys (total 5 of them) paid $6. What was the original cost of the new board game?	◯	◯	◯	◯
A shopkeeper buys 5 pens for $35 and sells them at the rate of $8 per pen. If he sells all the five pens, how much profit he will get?	◯	◯	◯	◯
Jeffrey bought 8 actions figures which cost 3 dollars each from John. John bought 6 books from the amount he received from Jeffrey. If the cost of each book John purchased is the same, what is the cost of each book?	◯	◯	◯	◯

Test ID	3M009
Test Name	Number Patterns
Student Name	
Date	

1	Ⓐ Ⓑ Ⓒ Ⓓ	6	Ⓐ Ⓑ Ⓒ Ⓓ	11	Ⓐ Ⓑ Ⓒ Ⓓ	16	Answer in the space provided below.
2	Ⓐ Ⓑ Ⓒ Ⓓ	7	Ⓐ Ⓑ Ⓒ Ⓓ	12	Ⓐ Ⓑ Ⓒ Ⓓ	17	Answer in the space provided below.
3	Ⓐ Ⓑ Ⓒ Ⓓ	8	Ⓐ Ⓑ Ⓒ Ⓓ	13	Ⓐ Ⓑ Ⓒ Ⓓ	18	Ⓐ Ⓑ Ⓒ Ⓓ
4	Ⓐ Ⓑ Ⓒ Ⓓ	9	Ⓐ Ⓑ Ⓒ Ⓓ	14	Ⓐ Ⓑ Ⓒ Ⓓ	19	Ⓐ Ⓑ Ⓒ Ⓓ
5	Ⓐ Ⓑ Ⓒ Ⓓ	10	Ⓐ Ⓑ Ⓒ Ⓓ	15	Ⓐ Ⓑ Ⓒ Ⓓ	20	Answer in the space provided below.

16

	10	35	24
2, 4, 6, 8	○	○	○
40, 36, 32, 28	○	○	○
7, 14, 21, 28	○	○	○

17

IN	OUT
3	9
4	
5	15
	18
7	

20

	2	4	5	7
A number has a four in its ones place. The number can be a multiple of ______.	○	○	○	○
A number has a five in its ones place. The number can be a multiple of ______.	○	○	○	○
A number has a zero in its ones place. The number can be a multiple of ______.	○	○	○	○
A number has a three in its ones place. The number can be a multiple of ______.	○	○	○	○

Date of Completion:_______________ Score:_______________

Chapter 1 → Lesson 9: Number Patterns

1. Which of the following is an even number?

   Ⓐ 764,723
   Ⓑ 90,835
   Ⓒ 5,862
   Ⓓ 609

2. Which of these sets contains no odd numbers?

   Ⓐ 13, 15, 81, 109, 199
   Ⓑ 123, 133, 421, 412, 600
   Ⓒ 34, 46, 48, 106, 88
   Ⓓ 12, 37, 6, 14, 144

3. Complete the following statement.
   The sum of two even numbers will always be _______ .

   Ⓐ greater than 10
   Ⓑ less than 100
   Ⓒ even
   Ⓓ odd

4. Complete the following statement.
   The product of two even numbers will always be ______ .

   Ⓐ even
   Ⓑ odd
   Ⓒ a multiple of 10
   Ⓓ a square number

5. Complete the following statement.
   A number has a nine in its ones place. The number must be a multiple of _____.

   Ⓐ 9
   Ⓑ 3
   Ⓒ 7
   Ⓓ None of the above

6. Complete the following statement.
Numbers that are multiples of 8 are all _______.

Ⓐ even
Ⓑ multiples of 2
Ⓒ multiples of 4
Ⓓ All of the above

7. If this pattern continues, what will the next 3 numbers be?
7, 14, 21, 28, 35, ...

Ⓐ 41, 47, 53
Ⓑ 49, 56, 63
Ⓒ 77, 84, 91
Ⓓ 42, 49, 56

8. Complete the following statement.
All multiples of _____ can be decomposed into two equal addends.

Ⓐ 6
Ⓑ 9
Ⓒ 3
Ⓓ 5

9. If this pattern continues, what will the next 3 numbers be?
9, 18, 27, 36, ...

Ⓐ 54, 63, 72
Ⓑ 45, 54, 63
Ⓒ 44, 52, 60
Ⓓ 44, 53, 62

10. A number is multiplied by 7 and the resulting product is even. Which of these could have been the number?

Ⓐ 7
Ⓑ 17
Ⓒ 34
Ⓓ 99

11. Complete the following statement.
The multiples of 4 will always ______.

Ⓐ have a 2 in the ones place
Ⓑ be even
Ⓒ be divisible by 8
Ⓓ None of these

12. Complete the following statement.
The sum of an even number and an odd number will always be _______.

Ⓐ even
Ⓑ odd
Ⓒ divisible by 3
Ⓓ None of the above

13. Complete the following statement.
A multiple of 4 can have a _____ in its ones place.

Ⓐ 2
Ⓑ 8
Ⓒ 6
Ⓓ All of the above

14. Complete the following statement.
A multiple of 5 can have a _____ as its ones digit.

Ⓐ 0
Ⓑ 3
Ⓒ 9
Ⓓ All of the above

15. Which of the following would produce an even product?

Ⓐ an even number times an even number
Ⓑ an even number times an odd number
Ⓒ an odd number times an even number
Ⓓ All of the above

16. Select the number that will come next if the pattern continues.

	10	35	24
2, 4, 6, 8	○	○	○
40, 36, 32, 28	○	○	○
7, 14, 21, 28	○	○	○

17. Type in the numbers that complete the table if the pattern is multiples of 3.

IN	OUT
3	9
4	
5	15
	18
7	

18. If the pattern continues, which of the following numbers will appear?
Note: More than one option may be correct.

100, 90, 80, 70

Ⓐ 50
Ⓑ 110
Ⓒ 80
Ⓓ 60

19. Complete the following statement. If you subtract an odd number from an even number, the difference will always be (a/an) _____. Circle the correct answer.

Ⓐ Multiple of 3
Ⓑ Even number
Ⓒ Odd number
Ⓓ Odd number or Even number

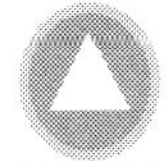

20. For each statement in the first column, choose all the correct answers.

	2	4	5	7
A number has a four in its ones place. The number can be a multiple of ______.	◯	◯	◯	◯
A number has a five in its ones place. The number can be a multiple of ______.	◯	◯	◯	◯
A number has a zero in its ones place. The number can be a multiple of ______.	◯	◯	◯	◯
A number has a three in its ones place. The number can be a multiple of ______.	◯	◯	◯	◯

End of Operations and Algebraic Thinking

Chapter 2
Number & Operations in Base Ten

Test ID	3M010
Test Name	Rounding Numbers
Student Name	
Date	

1	Ⓐ Ⓑ Ⓒ Ⓓ	6	Ⓐ Ⓑ Ⓒ Ⓓ	11	Ⓐ Ⓑ Ⓒ Ⓓ	16	Ⓐ Ⓑ Ⓒ Ⓓ
2	Ⓐ Ⓑ Ⓒ Ⓓ	7	Ⓐ Ⓑ Ⓒ Ⓓ	12	Ⓐ Ⓑ Ⓒ Ⓓ	17	Answer in the space provided below.
3	Ⓐ Ⓑ Ⓒ Ⓓ	8	Ⓐ Ⓑ Ⓒ Ⓓ	13	Ⓐ Ⓑ Ⓒ Ⓓ	18	Answer in the space provided below.
4	Ⓐ Ⓑ Ⓒ Ⓓ	9	Ⓐ Ⓑ Ⓒ Ⓓ	14	Ⓐ Ⓑ Ⓒ Ⓓ		
5	Ⓐ Ⓑ Ⓒ Ⓓ	10	Ⓐ Ⓑ Ⓒ Ⓓ	15	Ⓐ Ⓑ Ⓒ Ⓓ		

17

18

Number	Number when rounded to the nearest ten	Number when rounded to the nearest hundred
2,349	2,350	2,300
4,092		
8,396		

Date of Completion:_____________ Score:_____________ 

Lesson 1: Rounding Numbers

1. What is the value of the 9 in 11,291?

Ⓐ 9 ones
Ⓑ 9 hundreds
Ⓒ 9 thousands
Ⓓ 9 tens

2. What is the value of the digit 6 in 36,801?

Ⓐ Six thousand
Ⓑ Sixty
Ⓒ Sixty thousand
Ⓓ Six hundred

3. Which of these numbers has a 9 in the thousands place?

Ⓐ 690,099
Ⓑ 900
Ⓒ 209,866
Ⓓ 90,786

4. Round 2,564 to the nearest hundred.

Ⓐ 2,000
Ⓑ 2,500
Ⓒ 2,600
Ⓓ 2,700

5. Round 1,043 to the nearest hundred.

Ⓐ 1,000
Ⓑ 1,100
Ⓒ 1,040
Ⓓ 1,200

6. Round 537 to the nearest ten.

Ⓐ 500
Ⓑ 540
Ⓒ 550
Ⓓ 530

7. Round 957 to the nearest ten.

Ⓐ 960
Ⓑ 950
Ⓒ 900
Ⓓ 1,000

8. Maya is buying pencils for the school. Maya needs to buy enough pencils for 388 students. What is this number rounded to the nearest hundred?

Ⓐ 390
Ⓑ 380
Ⓒ 400
Ⓓ 500

9. Ninety-seven chairs are needed for an audience. What is this number rounded to the nearest ten?

Ⓐ 90
Ⓑ 100
Ⓒ 80
Ⓓ 110

10. Which of the following numbers does not round to 1,000 when rounding to the nearest hundred?

Ⓐ 955
Ⓑ 1,005
Ⓒ 1,051
Ⓓ 951

11. How many whole numbers, when rounded to the nearest ten give 100 as the result?

Ⓐ 8
Ⓑ 9
Ⓒ 10
Ⓓ 11

12. Fill in the blank.
795 rounds to 800 when rounded to the nearest ________.

Ⓐ ten
Ⓑ hundred
Ⓒ ten or hundred
Ⓓ thousand

13. Fill in the blank.
1,090 rounds to 1,100 when rounded to the nearest ________.

Ⓐ ten
Ⓑ hundred
Ⓒ ten or hundred
Ⓓ thousand

14. The attendance at a local baseball game is announced to be 4,328. What is this number rounded to the nearest ten?

Ⓐ 4,300
Ⓑ 4,330
Ⓒ 4,320
Ⓓ 4,400

15. The number of plants in a garden, when rounded to the nearest hundred, rounds to 800. Which of the following could not be the number of plants in the garden?

Ⓐ 850
Ⓑ 800
Ⓒ 750
Ⓓ 849

16. Which numbers represent the number 617 when rounded to the nearest ten or hundred? Circle all correct answers.

Ⓐ 620
Ⓑ 600
Ⓒ 700
Ⓓ 630

17. Round 489 to the nearest hundred. Write the correct answer into the box.

18. Complete the table in the format given in the example.

Number	Number when rounded to the nearest ten	Number when rounded to the nearest hundred
2,349	2,350	2,300
4,092		
8,396		

Test ID	3M011
Test Name	Addition & Subtraction
Student Name	
Date	

1	Ⓐ Ⓑ Ⓒ Ⓓ	6	Ⓐ Ⓑ Ⓒ Ⓓ	11	Ⓐ Ⓑ Ⓒ Ⓓ	16	Answer in the space provided below.
2	Ⓐ Ⓑ Ⓒ Ⓓ	7	Ⓐ Ⓑ Ⓒ Ⓓ	12	Ⓐ Ⓑ Ⓒ Ⓓ	17	Answer in the space provided below.
3	Ⓐ Ⓑ Ⓒ Ⓓ	8	Ⓐ Ⓑ Ⓒ Ⓓ	13	Ⓐ Ⓑ Ⓒ Ⓓ	18	Answer in the space provided below.
4	Ⓐ Ⓑ Ⓒ Ⓓ	9	Ⓐ Ⓑ Ⓒ Ⓓ	14	Ⓐ Ⓑ Ⓒ Ⓓ	19	Answer in the space provided below.
5	Ⓐ Ⓑ Ⓒ Ⓓ	10	Ⓐ Ⓑ Ⓒ Ⓓ	15	Ⓐ Ⓑ Ⓒ Ⓓ		

16

	899	467	558
422 + 136	○	○	○
608 + 291	○	○	○
157 + 310	○	○	○

17

18

	Hundreds	Tens	Ones
	2		5
+		3	
Total	8	4	9

19

Date of Completion:_______________ Score:_______________

Chapter 2 → Lesson 2: Addition & Subtraction

1. What is the standard form of 70,000 + 6,000 + 800 + 60 + 2?

Ⓐ 706,862
Ⓑ 76,862
Ⓒ 7,682
Ⓓ 782

2. Two numbers have a difference of 29. The two numbers could be ____________.

Ⓐ 11 and 18
Ⓑ 23 and 42
Ⓒ 40 and 11
Ⓓ 50 and 39

3. Two numbers add up to 756. One number is 356. What is the other number?

Ⓐ 356
Ⓑ 300
Ⓒ 400
Ⓓ 456

4. Which of these expressions has the same difference as 94 - 50?

Ⓐ 70 - 34
Ⓑ 80 - 46
Ⓒ 60 - 16
Ⓓ 90 - 54

5. Which of these number sentences is not true?

Ⓐ 88 + 12 = 90 + 10
Ⓑ 82 + 18 = 88 + 12
Ⓒ 56 + 45 = 54 + 56
Ⓓ 46 + 15 = 56 + 5

6. Jim has 640 baseball cards and 280 basketball cards. How many sports cards does Jim have in all?

   Ⓐ 820 cards
   Ⓑ 360 cards
   Ⓒ 8,120 cards
   Ⓓ 920 cards

7. Find the difference.
   860 - 659

   Ⓐ 219
   Ⓑ 319
   Ⓒ 201
   Ⓓ 19

8. The students made 565 book covers for their math books. They used up 422 of the book covers. How many book covers are left?

   Ⓐ 242 book covers
   Ⓑ 163 book covers
   Ⓒ 987 book covers
   Ⓓ 143 book covers

9. What is the difference of 32 and 5?

   Ⓐ 33
   Ⓑ 27
   Ⓒ 160
   Ⓓ 37

10. Jenny plans to sell 50 boxes of cookies to help her scout troop raise funds. She sold 20 boxes to her neighbors. Her dad sold 15 boxes at his work office. How many more boxes does she need to sell to meet her goal?

   Ⓐ 15 boxes
   Ⓑ 10 boxes
   Ⓒ 5 boxes
   Ⓓ 25 boxes

11. Sara had 124 stickers. She gave away 62 stickers and bought 73 more stickers. How many stickers does Sara have now?

   Ⓐ 135 stickers
   Ⓑ 62 stickers
   Ⓒ 120 stickers
   Ⓓ 11 stickers

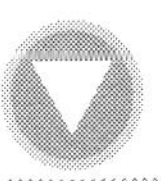

12. Which of these addition expressions would require regrouping of hundreds and tens?

Ⓐ 923 + 37
Ⓑ 456 + 443
Ⓒ 235 + 234
Ⓓ 576 + 442

13. There were 605 people sitting in an auditorium at the start of a show. Thirty-five people left during the intermission. How many people remained in the auditorium after the intermission?

Ⓐ 630 people
Ⓑ 580 people
Ⓒ 570 people
Ⓓ 595 people

14. If 3 tens are subtracted from 401, what is the difference?

Ⓐ 471
Ⓑ 371
Ⓒ 398
Ⓓ 381

15. Find the sum of 37 + 93 + 200.

Ⓐ 330
Ⓑ 663
Ⓒ 320
Ⓓ 300

16. Select the correct sum for each addition expression.

	899	467	558
422 + 136	○	○	○
608 + 291	○	○	○
157 + 310	○	○	○

17. Hannah received a score of 604 on the exam. Ben received a score of 719. What was the difference between the two scores? Write your answer in the box given below.

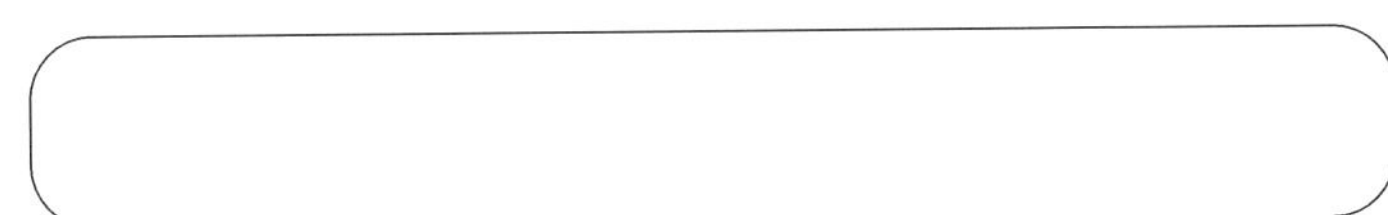

18. Type in the correct numbers to make the sum true.

	Hundreds	Tens	Ones
	2		5
+		3	
Total	8	4	9

19. Karen has 805 milliliters of milk. After he drinks some milk, 538 milliliters are left. How much milk did Karen drink? Show the steps by which you arrive at the answer.

Test ID	3M012
Test Name	Multiplying Multiples of 10
Student Name	
Date	

1	(A) (B) (C) (D)	6	(A) (B) (C) (D)	11	(A) (B) (C) (D)	16	(A) (B) (C) (D)
2	(A) (B) (C) (D)	7	(A) (B) (C) (D)	12	(A) (B) (C) (D)	17	(A) (B) (C)
3	(A) (B) (C) (D)	8	(A) (B) (C) (D)	13	(A) (B) (C) (D)	18	Answer in the space provided below.
4	(A) (B) (C) (D)	9	(A) (B) (C) (D)	14	(A) (B) (C) (D)	19	Answer in the space provided below.
5	(A) (B) (C) (D)	10	(A) (B) (C) (D)	15	(A) (B) (C) (D)	20	Answer in the space provided below.

18

19

6 x 50 =	
8 x 60 =	
70 x 8 =	
80 x 9 =	

20

	630	320	540	450
9 x 60 =	○	○	○	○
4 x 80 =	○	○	○	○
90 x 5 =	○	○	○	○
90 x 7 =	○	○	○	○

Date of Completion:_______________ Score:_______________

Chapter 2 → Lesson 3: Multiplying Multiples of 10

1. **Multiply:**
   6 x 10 = ____

   Ⓐ 66
   Ⓑ 60
   Ⓒ 61
   Ⓓ 16

2. **What is the product of 10 and 10?**

   Ⓐ 20
   Ⓑ 50
   Ⓒ 100
   Ⓓ 1,000

3. **Find the product.**
   5 x 40 = _____

   Ⓐ 100
   Ⓑ 90
   Ⓒ 200
   Ⓓ 240

4. **Multiply:**
   ____ = 6 x 60

   Ⓐ 120
   Ⓑ 180
   Ⓒ 320
   Ⓓ 360

5. **Find the product of 70 and 7.**

   Ⓐ 77
   Ⓑ 140
   Ⓒ 420
   Ⓓ 490

6. Multiply:
   _____ = 30 x 7

   Ⓐ 210
   Ⓑ 240
   Ⓒ 180
   Ⓓ 100

7. Find the product.
   90 x 9 = ____

   Ⓐ 800
   Ⓑ 810
   Ⓒ 900
   Ⓓ 1,800

8. Multiply:
   8 x 80 = _____

   Ⓐ 160
   Ⓑ 620
   Ⓒ 640
   Ⓓ 660

9. Find the product.
   2 x 70 = ____

   Ⓐ 140
   Ⓑ 120
   Ⓒ 90
   Ⓓ 160

10. Multiply:
    90 x 3 = ____

    Ⓐ 120
    Ⓑ 180
    Ⓒ 270
    Ⓓ 290

11. **Multiply:**
2 x 10 = ____

Ⓐ **2**
Ⓑ **5**
Ⓒ **10**
Ⓓ **20**

12. **Multiply:**
50 x 1 = ____

Ⓐ **1**
Ⓑ **25**
Ⓒ **49**
Ⓓ **50**

13. **Multiply:**
10 x 3 = _____

Ⓐ **10**
Ⓑ **30**
Ⓒ **100**
Ⓓ **300**

14. **Find the product of 30 and 9.**

Ⓐ **270**
Ⓑ **39**
Ⓒ **3**
Ⓓ **100**

15. **Multiply:**
10 x 7 = ____

Ⓐ **70**
Ⓑ **10**
Ⓒ **100**
Ⓓ **7**

16. Select the math sentences that can be used to find the product of 70 and 5. Select all the relevant math sentences.

Ⓐ 7 x 5 = 35
Ⓑ 35 x 10= 350
Ⓒ 70 x 50= 3,500
Ⓓ 70 x 5= 350

17. Circle the picture that shows the product of 20 and 2.

Instruction: ☐ = 10

Ⓐ

Ⓑ

Ⓒ

18. Find the product of 30 x 8. Write your answer in the box below.

19. Complete the following table:

6 x 50 =	
8 x 60 =	
70 x 8 =	
80 x 9 =	

20. Match the multiplication expression with the correct product.

	630	320	540	450
9 x 60 =				
4 x 80 =				
90 x 5 =				
90 x 7 =				

End of Number & Operations in Base Ten

Chapter 3

Number & Operations - Fractions

Test ID	3M013
Test Name	Fractions of a Whole
Student Name	
Date	

1	Ⓐ Ⓑ Ⓒ Ⓓ	6	Ⓐ Ⓑ Ⓒ Ⓓ	11	Ⓐ Ⓑ Ⓒ Ⓓ	16	Answer in the space provided below.
2	Ⓐ Ⓑ Ⓒ Ⓓ	7	Ⓐ Ⓑ Ⓒ Ⓓ	12	Ⓐ Ⓑ Ⓒ Ⓓ	17	Answer in the space provided below.
3	Ⓐ Ⓑ Ⓒ Ⓓ	8	Ⓐ Ⓑ Ⓒ Ⓓ	13	Ⓐ Ⓑ Ⓒ Ⓓ	18	Ⓐ Ⓑ Ⓒ Ⓓ
4	Ⓐ Ⓑ Ⓒ Ⓓ	9	Ⓐ Ⓑ Ⓒ Ⓓ	14	Ⓐ Ⓑ Ⓒ Ⓓ		
5	Ⓐ Ⓑ Ⓒ Ⓓ	10	Ⓐ Ⓑ Ⓒ Ⓓ	15	Ⓐ Ⓑ Ⓒ Ⓓ		

16

	Yes	No
1/8	○	○
1/4	○	○
1/3	○	○

17

Figure	Fraction
■□□□□□□□	
■□□□□	
□■	

Lesson 1: Fractions of a Whole

1. What fraction of the letters in the word "READING" are vowels?

   Ⓐ $\frac{4}{7}$

   Ⓑ $\frac{3}{4}$

   Ⓒ $\frac{3}{7}$

   Ⓓ $\frac{1}{3}$

2. A bag contains 3 red, 2 yellow, and 5 blue tiles. What fraction of the tiles are yellow?

   Ⓐ $\frac{2}{5}$

   Ⓑ $\frac{2}{10}$

   Ⓒ $\frac{3}{7}$

   Ⓓ $\frac{1}{3}$

3. A rectangle is cut into four equal pieces. Each piece represents what fraction of the rectangle?

   Ⓐ one half
   Ⓑ one third
   Ⓒ one fourth
   Ⓓ one fifth

4. What fraction of the square is shaded?

Ⓐ $\frac{1}{2}$

Ⓑ $\frac{1}{3}$

Ⓒ $\frac{2}{1}$

Ⓓ $\frac{1}{1}$

5. What fraction of the square is shaded?

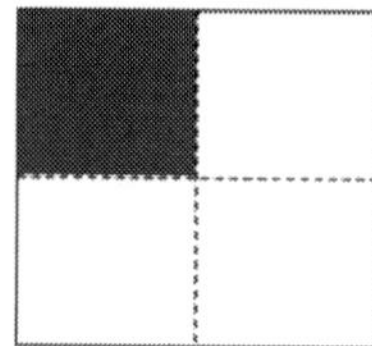

Ⓐ $\frac{1}{2}$

Ⓑ $\frac{1}{4}$

Ⓒ $\frac{1}{3}$

Ⓓ $\frac{3}{1}$

6. **What fraction of the square is NOT shaded?**

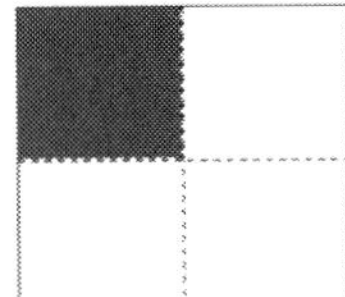

Ⓐ $\frac{1}{2}$

Ⓑ $\frac{1}{4}$

Ⓒ $\frac{3}{1}$

Ⓓ $\frac{3}{4}$

7. **What fraction of the circle is shaded?**

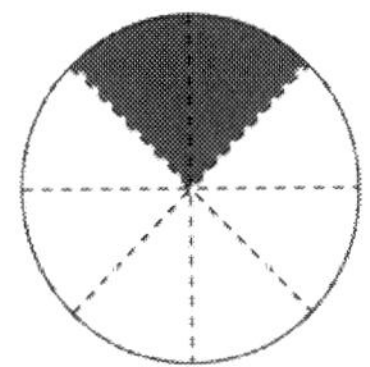

Ⓐ $\frac{1}{8}$

Ⓑ $\frac{2}{8}$

Ⓒ $\frac{2}{6}$

Ⓓ $\frac{6}{2}$

8. **What fraction of the circle is not shaded?**

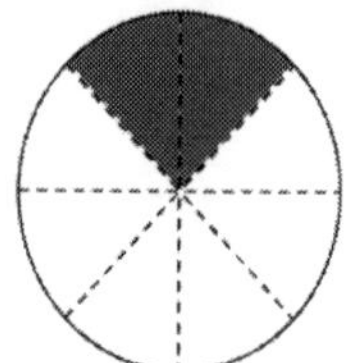

Ⓐ $\frac{6}{8}$

Ⓑ $\frac{7}{8}$

Ⓒ $\frac{2}{6}$

Ⓓ $\frac{6}{2}$

9. **What fraction of the circle is shaded?**

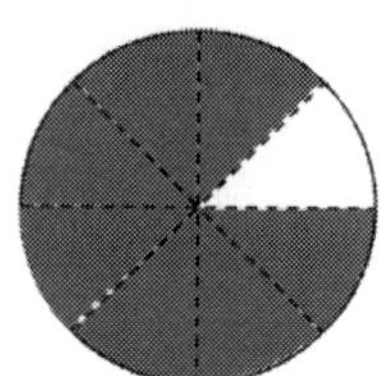

Ⓐ $\frac{1}{8}$

Ⓑ $\frac{1}{7}$

Ⓒ $\frac{7}{1}$

Ⓓ $\frac{7}{8}$

10. What fraction of the circle is not shaded?

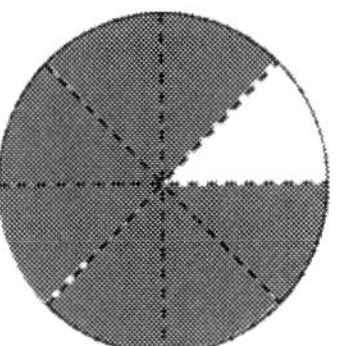

Ⓐ $\frac{1}{8}$

Ⓑ $\frac{1}{7}$

Ⓒ $\frac{7}{1}$

Ⓓ $\frac{8}{1}$

11. What fraction of the rectangle is shaded?

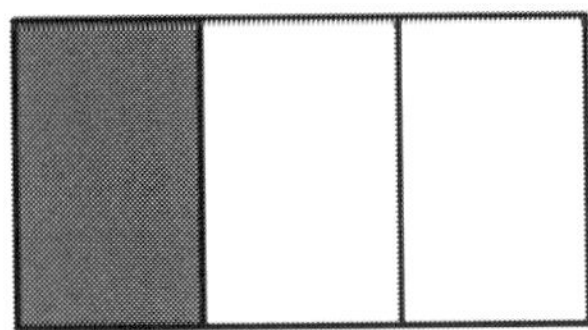

Ⓐ $\frac{1}{2}$

Ⓑ $\frac{1}{3}$

Ⓒ $\frac{2}{3}$

Ⓓ $\frac{2}{1}$

12. What fraction of the rectangle is not shaded?

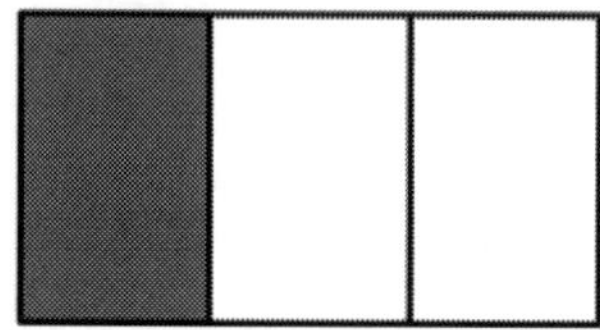

Ⓐ $\frac{1}{3}$

Ⓑ $\frac{1}{2}$

Ⓒ $\frac{2}{3}$

Ⓓ $\frac{2}{1}$

13. A pizza is cut into 12 equal slices. Eight slices are eaten. What fraction of the pizza is left?

Ⓐ $\frac{8}{12}$

Ⓑ $\frac{4}{8}$

Ⓒ $\frac{4}{12}$

Ⓓ $\frac{8}{4}$

14. The class has 20 children. Only half of the students brought their homework. How many students have their homework?

Ⓐ 20 students
Ⓑ 15 students
Ⓒ 10 students
Ⓓ 12 students

15. Meagan has 24 cupcakes. She gives a third of them to Micah. How many cupcakes does Micah have?

Ⓐ 8 cupcakes
Ⓑ 12 cupcakes
Ⓒ 3 cupcakes
Ⓓ 4 cupcakes

16. Which of the following fractions could apply to this figure? Complete the table by selecting yes or no.

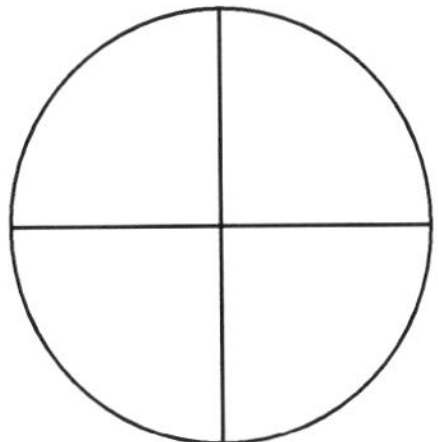

	Yes	No
1/8	◯	◯
1/4	◯	◯
1/3	◯	◯

17. What fraction does each figure show? Write your answers in the blank boxes.

Figure	Fraction

18. Which of the following fractions could apply to this figure? Select all correct answers.

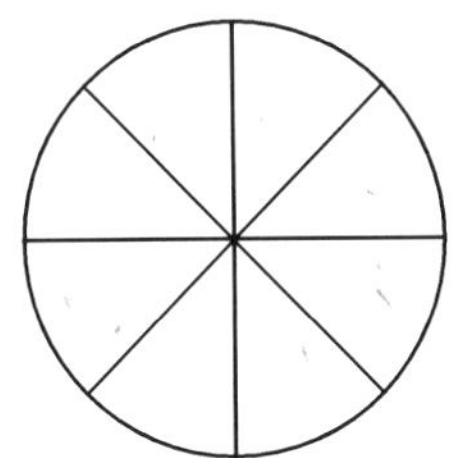

Ⓐ $\frac{1}{3}$

Ⓑ $\frac{1}{8}$

Ⓒ $\frac{1}{5}$

Ⓓ $\frac{8}{8}$

Test ID	3M014
Test Name	Fractions on the Number Line
Student Name	
Date	

1	Ⓐ Ⓑ Ⓒ Ⓓ	6	Ⓐ Ⓑ Ⓒ Ⓓ	11	Ⓐ Ⓑ Ⓒ Ⓓ	16	Ⓐ Ⓑ Ⓒ Ⓓ
2	Ⓐ Ⓑ Ⓒ Ⓓ	7	Ⓐ Ⓑ Ⓒ Ⓓ	12	Ⓐ Ⓑ Ⓒ Ⓓ	17	Answer in the space provided below.
3	Ⓐ Ⓑ Ⓒ Ⓓ	8	Ⓐ Ⓑ Ⓒ Ⓓ	13	Ⓐ Ⓑ Ⓒ Ⓓ	18	Answer in the space provided below.
4	Ⓐ Ⓑ Ⓒ Ⓓ	9	Ⓐ Ⓑ Ⓒ Ⓓ	14	Ⓐ Ⓑ Ⓒ Ⓓ	19	Answer in the space provided below.
5	Ⓐ Ⓑ Ⓒ Ⓓ	10	Ⓐ Ⓑ Ⓒ Ⓓ	15	Ⓐ Ⓑ Ⓒ Ⓓ		

17

18

19

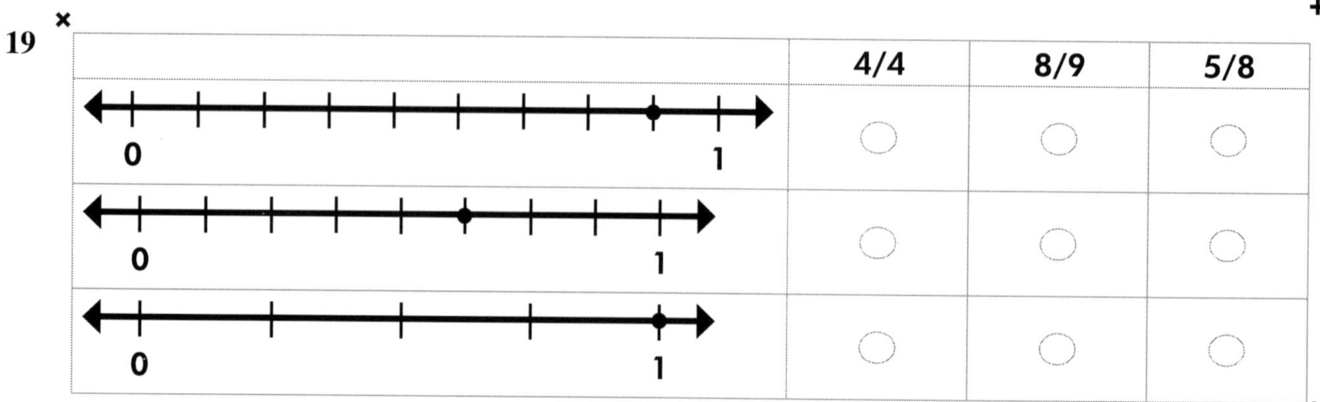

Chapter 3 → Lesson 2: Fractions on the Number Line

1. What fraction does the dot marked on the number line indicate?

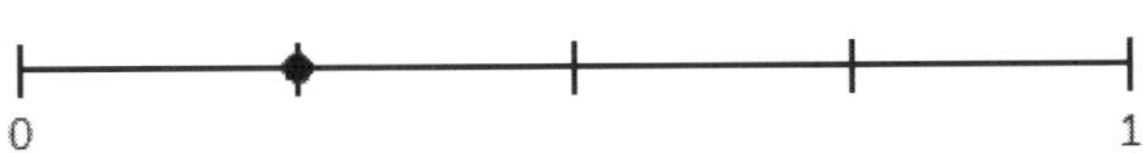

Ⓐ $\frac{1}{4}$

Ⓑ $\frac{1}{3}$

Ⓒ $\frac{3}{4}$

Ⓓ $\frac{4}{4}$

2. What fraction does the dot marked on the number line show?

Ⓐ $\frac{1}{2}$

Ⓑ $\frac{2}{2}$

Ⓒ $\frac{1}{3}$

Ⓓ $\frac{2}{3}$

3. **What fraction does the dot marked on the number line show?**

Ⓐ $\frac{2}{8}$

Ⓑ $\frac{3}{5}$

Ⓒ $\frac{3}{8}$

Ⓓ $\frac{4}{8}$

4. **What fraction does the dot marked on the number line show?**

Ⓐ $\frac{3}{8}$

Ⓑ $\frac{6}{8}$

Ⓒ $\frac{5}{8}$

Ⓓ $\frac{4}{8}$

5. **What fraction does the dot marked on the number line show?**

Ⓐ $\frac{1}{6}$

Ⓑ $\frac{4}{6}$

Ⓒ $\frac{3}{6}$

Ⓓ $\frac{1}{5}$

6. **What fraction does the dot marked on the number line show?**

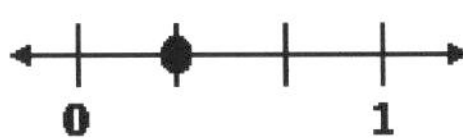

Ⓐ $\frac{2}{4}$

Ⓑ $\frac{2}{3}$

Ⓒ $\frac{1}{3}$

Ⓓ $\frac{1}{4}$

7. **What fraction does the dot marked on the number line show?**

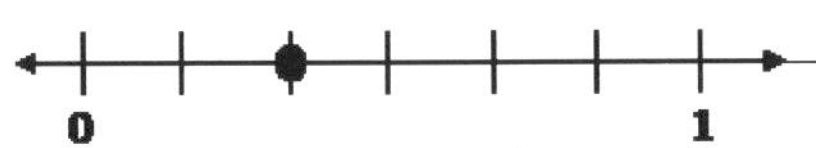

Ⓐ $\frac{1}{6}$

Ⓑ $\frac{3}{6}$

Ⓒ $\frac{2}{4}$

Ⓓ $\frac{2}{6}$

8. **What fraction does the dot marked on the number line show?**

Ⓐ $\frac{4}{8}$

Ⓑ $\frac{5}{8}$

Ⓒ $\frac{4}{4}$

Ⓓ $\frac{2}{8}$

9. What fraction does the dot marked on the number line show?

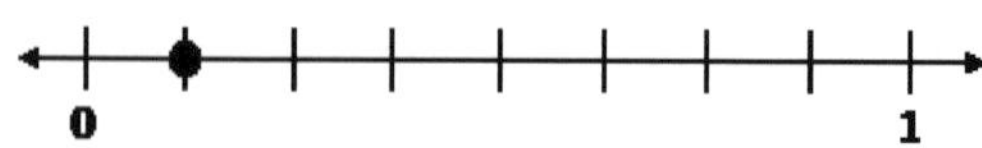

Ⓐ $\frac{2}{8}$

Ⓑ $\frac{2}{9}$

Ⓒ $\frac{1}{9}$

Ⓓ $\frac{1}{8}$

10. What fraction does the dot marked on the number line show

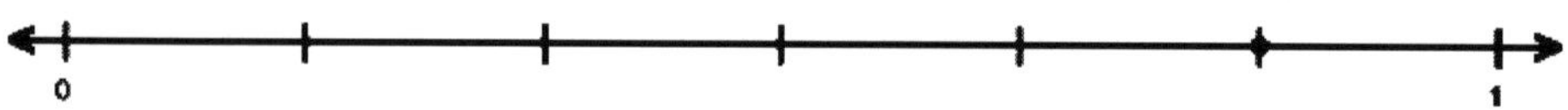

Ⓐ $\frac{4}{6}$

Ⓑ $\frac{5}{6}$

Ⓒ $\frac{3}{6}$

Ⓓ $\frac{1}{6}$

11. What fraction does the dot marked on the number line show?

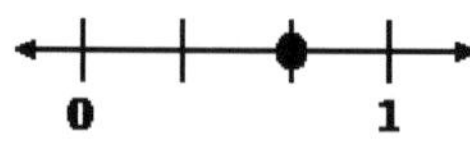

Ⓐ $\frac{2}{3}$

Ⓑ $\frac{1}{3}$

Ⓒ $\frac{3}{3}$

Ⓓ $\frac{4}{3}$

12. What fraction does the dot marked on the number line show?

Ⓐ $\frac{1}{4}$

Ⓑ $\frac{2}{4}$

Ⓒ $\frac{3}{4}$

Ⓓ $\frac{4}{4}$

13. What fraction does the dot marked on the number line show?

Ⓐ $\frac{8}{9}$

Ⓑ $\frac{7}{8}$

Ⓒ $\frac{2}{8}$

Ⓓ $\frac{5}{9}$

14. What fraction does the dot marked on the number line show?

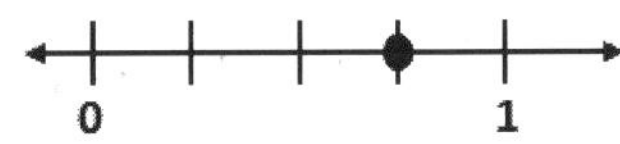

Ⓐ $\frac{1}{4}$

Ⓑ $\frac{3}{3}$

Ⓒ $\frac{2}{4}$

Ⓓ $\frac{3}{4}$

15. What fraction does v the dot marked on the number line show?

0 1

Ⓐ $\frac{7}{8}$

Ⓑ $\frac{6}{8}$

Ⓒ $\frac{5}{8}$

Ⓓ $\frac{5}{3}$

16. Which fractions does the dot marked on the number line show? Select all correct answers.

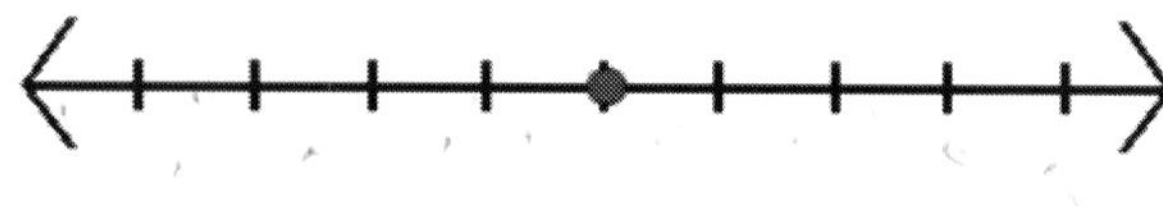

0 1

Ⓐ $\frac{3}{6}$

Ⓑ $\frac{1}{4}$

Ⓒ $\frac{4}{8}$

Ⓓ $\frac{1}{2}$

17. What fraction does the dot marked on the number line show? Write your answer in the box below.

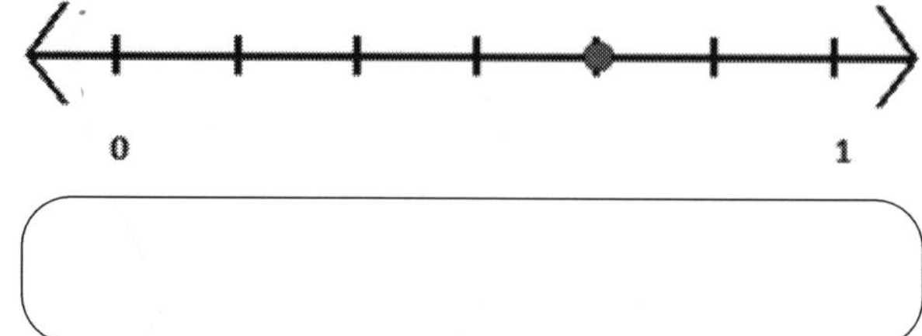

18. Draw a number line and locate the fraction $\frac{5}{7}$ on it.

19. There are 3 number lines in the first column. Which fractions are represented by the dots on the number lines? For each number line, select the correct answer.

	4/4	8/9	5/8
0 1	○	○	○
0 1	○	○	○
0 1	○	○	○

Test ID	3M015
Test Name	Comparing Fractions
Student Name	
Date	

1	Ⓐ Ⓑ Ⓒ Ⓓ	6	Ⓐ Ⓑ Ⓒ Ⓓ	11	Ⓐ Ⓑ Ⓒ Ⓓ	16	Answer in the space provided below.
2	Ⓐ Ⓑ Ⓒ Ⓓ	7	Ⓐ Ⓑ Ⓒ Ⓓ	12	Ⓐ Ⓑ Ⓒ Ⓓ	17	Answer in the space provided below.
3	Ⓐ Ⓑ Ⓒ Ⓓ	8	Ⓐ Ⓑ Ⓒ Ⓓ	13	Ⓐ Ⓑ Ⓒ Ⓓ	18	Ⓐ Ⓑ Ⓒ Ⓓ
4	Ⓐ Ⓑ Ⓒ Ⓓ	9	Ⓐ Ⓑ Ⓒ Ⓓ	14	Ⓐ Ⓑ Ⓒ Ⓓ	19	Ⓐ Ⓑ Ⓒ Ⓓ
5	Ⓐ Ⓑ Ⓒ Ⓓ	10	Ⓐ Ⓑ Ⓒ Ⓓ	15	Ⓐ Ⓑ Ⓒ Ⓓ	20	Answer in the space provided below.

16

	Yes	No
$\frac{1}{4}$	○	○
$\frac{4}{4}$	○	○
$\frac{2}{4}$	○	○

17

Fraction #1	< = or >	Fraction #2
$\frac{2}{3}$		$\frac{1}{3}$
$\frac{4}{6}$		$\frac{5}{6}$
$\frac{8}{8}$		$\frac{5}{5}$
$\frac{3}{4}$		$\frac{1}{4}$

20

Chapter 3 → Lesson 3: Comparing Fractions

1. Which of these sets has the fractions listed from least to greatest?

Ⓐ $\frac{1}{6}, \frac{1}{4}, \frac{1}{3}, \frac{1}{2}$

Ⓑ $\frac{1}{2}, \frac{1}{3}, \frac{1}{6}, \frac{1}{4}$

Ⓒ $\frac{1}{3}, \frac{1}{4}, \frac{1}{2}, \frac{1}{6}$

Ⓓ $\frac{1}{2}, \frac{1}{3}, \frac{1}{4}, \frac{1}{6}$

2. Which of these fractions would be found between $\frac{1}{2}$ and 1 on a number line?

Ⓐ $\frac{1}{4}$

Ⓑ $\frac{1}{3}$

Ⓒ $\frac{5}{8}$

Ⓓ $\frac{3}{1}$

3. Which of these fractions would be found between 0 an $\frac{1}{2}$ on a number line?

Ⓐ $\frac{7}{8}$

Ⓑ $\frac{3}{4}$

Ⓒ $\frac{1}{4}$

Ⓓ $\frac{5}{8}$

4. Which of these fractions would be found between 0 and $\frac{3}{4}$ on a number line?

Ⓐ $\frac{7}{8}$

Ⓑ $\frac{4}{8}$

Ⓒ $\frac{5}{6}$

Ⓓ $\frac{4}{4}$

5. Which of these fractions is less than $\frac{6}{8}$?

Ⓐ $\frac{1}{8}$

Ⓑ $\frac{7}{8}$

Ⓒ $\frac{9}{8}$

Ⓓ $\frac{8}{8}$

6. Answer the following: $\frac{1}{2}$ > ________?

Ⓐ $\frac{1}{4}$

Ⓑ $\frac{2}{3}$

Ⓒ $\frac{4}{8}$

Ⓓ $\frac{2}{2}$

7. Which fraction is greater: $\frac{4}{8}$ or $\frac{1}{2}$?

Ⓐ $\frac{1}{2}$

Ⓑ $\frac{4}{8}$

Ⓒ They are equal.

Ⓓ There is not enough information given.

8. Which fraction is less: $\frac{1}{4}$ or $\frac{1}{8}$?

Ⓐ $\frac{1}{4}$

Ⓑ $\frac{1}{8}$

Ⓒ They are equal.

Ⓓ There is not enough information given.

9. Which fraction is less $\frac{4}{6}$ or $\frac{1}{6}$?

Ⓐ $\frac{4}{6}$

Ⓑ $\frac{1}{6}$

Ⓒ They are equal.

Ⓓ There is not enough information given.

10. If two fractions have the same denominator, the one with a(n) __________ numerator is the greater fraction.

Ⓐ smaller
Ⓑ greater
Ⓒ even
Ⓓ zero

11. Complete this number sentence with <, > or = +: $\frac{4}{6}$ _____ $\frac{5}{6}$

Ⓐ >
Ⓑ <
Ⓒ =
Ⓓ There is not enough information given.

12. Complete this number sentence: $\frac{3}{8}$ _____ $\frac{5}{8}$

Ⓐ >
Ⓑ <
Ⓒ =
Ⓓ There is not enough information given.

13. Complete this number sentence: $\frac{2}{4}$ ______ $\frac{1}{2}$

Ⓐ >
Ⓑ <
Ⓒ =
Ⓓ There is not enough information given.

14. Complete this number sentence: $\frac{1}{2}$ ______ $\frac{1}{3}$

Ⓐ >
Ⓑ <
Ⓒ =
Ⓓ There is not enough information given.

15. To compare fractions with the same numerator, you need to look at the ___________.

Ⓐ numerators
Ⓑ denominators
Ⓒ factors of the numerator
Ⓓ multiples of the denominator

16. Are the following fractions less than $\frac{3}{4}$? Select yes or no.

	Yes	No
$\frac{1}{4}$	○	○
$\frac{4}{4}$	○	○
$\frac{2}{4}$	○	○

17. Is Fraction #1 less than, greater than, or equal to Fraction #2? Write the correct symbol in the empty boxes.

Fraction #1	< = or >	Fraction #2
$\frac{2}{3}$		$\frac{1}{3}$
$\frac{4}{6}$		$\frac{5}{6}$
$\frac{8}{8}$		$\frac{5}{5}$
$\frac{3}{4}$		$\frac{1}{4}$

18. Which of the following fractions are greater than $\frac{2}{5}$? Select all correct answers.

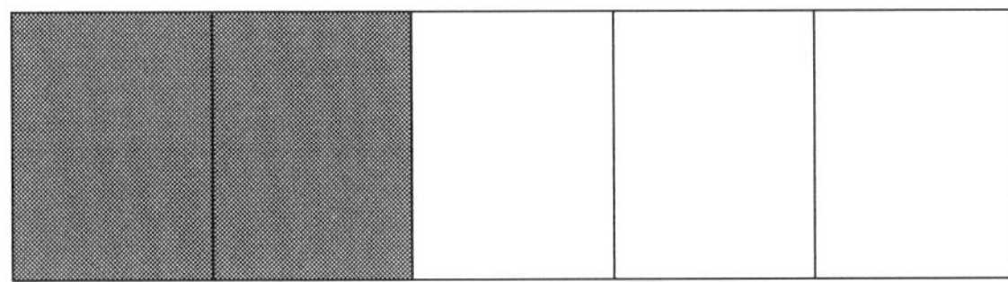

Ⓐ $\frac{1}{5}$

Ⓑ $\frac{3}{5}$

Ⓒ $\frac{4}{5}$

Ⓓ $\frac{2}{5}$

19. Which of these fractions is greater than $\frac{5}{7}$? Circle the correct answer.

Ⓐ $\frac{1}{7}$

Ⓑ $\frac{2}{7}$

Ⓒ $\frac{4}{7}$

Ⓓ $\frac{6}{7}$

20. Which of the following fractions is the least?

$\frac{6}{5}, \frac{6}{4}, \frac{6}{10}, \frac{6}{9}.$

Write your answer in the box below.

End of Number & Operations - Fractions

Chapter 4

Measurement and Data

Test ID	3M016
Test Name	Telling Time
Student Name	
Date	

1	Ⓐ Ⓑ Ⓒ Ⓓ	6	Ⓐ Ⓑ Ⓒ Ⓓ	11	Ⓐ Ⓑ Ⓒ Ⓓ	16	Ⓐ Ⓑ Ⓒ Ⓓ
2	Ⓐ Ⓑ Ⓒ Ⓓ	7	Ⓐ Ⓑ Ⓒ Ⓓ	12	Ⓐ Ⓑ Ⓒ Ⓓ	17	Answer in the space provided below.
3	Ⓐ Ⓑ Ⓒ Ⓓ	8	Ⓐ Ⓑ Ⓒ Ⓓ	13	Ⓐ Ⓑ Ⓒ Ⓓ	18	Answer in the space provided below.
4	Ⓐ Ⓑ Ⓒ Ⓓ	9	Ⓐ Ⓑ Ⓒ Ⓓ	14	Ⓐ Ⓑ Ⓒ Ⓓ	19	Answer in the space provided below.
5	Ⓐ Ⓑ Ⓒ Ⓓ	10	Ⓐ Ⓑ Ⓒ Ⓓ	15	Ⓐ Ⓑ Ⓒ Ⓓ	20	Answer in the space provided below.

17

18

19

20

9:42	11:58	2:03
○	○	○
○	○	○
○	○	○

Date of Completion:______________ Score:______________

Lesson 1: Telling Time

1. **What time does this clock show?**

Ⓐ **3:12**
Ⓑ **2:17**
Ⓒ **2:22**
Ⓓ **2:03**

2. **What time does this clock show?**

Ⓐ **5:42**
Ⓑ **9:28**
Ⓒ **6:47**
Ⓓ **5:47**

3. **What time does this clock show?**

Ⓐ **10:00**
Ⓑ **12:50**
Ⓒ **10:02**
Ⓓ **9:41**

4. What time does this clock show?

Ⓐ 12:39
Ⓑ 8:04
Ⓒ 1:38
Ⓓ 12:42

5. On an analog clock, the shorter hand shows the _____ .

Ⓐ minutes
Ⓑ hours
Ⓒ seconds
Ⓓ days

6. On an analog clock, the longer hand shows the ______ .

Ⓐ minutes
Ⓑ hours
Ⓒ days
Ⓓ seconds

7. The clock currently shows:

What time will it be in 8 minutes?

Ⓐ 1:38
Ⓑ 10:15
Ⓒ 10:10
Ⓓ 12:58

8. The clock currently shows:

What time will it be in 20 minutes?

Ⓐ 12:59
Ⓑ 1:09
Ⓒ 2:00
Ⓓ 8:24

9. The clock says:

What time was it 10 minutes ago?

Ⓐ 1:29
Ⓑ 12:29
Ⓒ 12:49
Ⓓ 1:09

10. Lucy started her test at 12:09 PM and finished at 12:58 PM. David started at 12:15 PM and ended at 1:03 PM. Who finished in a shorter amount of time?

Ⓐ Lucy
Ⓑ David
Ⓒ They both took the same amount of time.
Ⓓ There is not enough information given.

11. The Jamisons are on a road trip that will take 5 hours and 25 minutes. They have been driving for 3 hours and 41 minutes. How much longer do they need to travel before they reach their destination?

Ⓐ 1 hour, 13 minutes
Ⓑ 2 hours, 19 minutes
Ⓒ 1 hour, 44 minutes
Ⓓ 2 hours, 7 minutes

12. Rachel usually gets around 9 hours of sleep per night. She went to bed at 9:30 PM. About what time will she wake up?

Ⓐ 8:30 AM
Ⓑ 10:30 AM
Ⓒ 6:30 AM
Ⓓ 5:30 AM

13. A 45 minute long show ends at 12:20 PM. When did the show begin?

Ⓐ 1:05 PM
Ⓑ 11:35 AM
Ⓒ 11:35 PM
Ⓓ 11:45 AM

14. Mrs. James is giving her class a math test. She is allowing the students 40 minutes to finish the test. The test began at 10:22 AM. By what time must the test be finished?

Ⓐ 10:42 AM
Ⓑ 10:57 AM
Ⓒ 11:02 AM
Ⓓ 12:02 PM

15. The directions on a frozen pizza say to cook it for 25 minutes. Mr. Adams puts the frozen pizza in the oven at 5:43 PM. When will the pizza be done?

Ⓐ 6:08 PM
Ⓑ 6:18 PM
Ⓒ 6:13 PM
Ⓓ 5:58 PM

16. Which statements are true? Select all the correct answers.

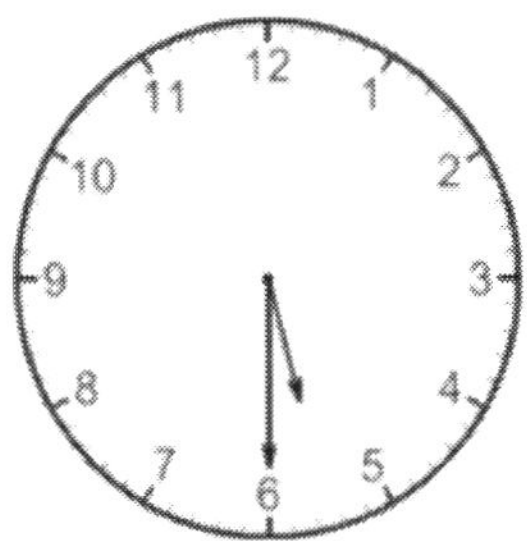

Ⓐ The minute hand points to 4
Ⓑ The minute hand points to 6
Ⓒ The hour hand points to 6
Ⓓ The clock shows the time as 5:30

17. What time does this clock show? Write your answer in the box below.

18. Circle the clock that shows the time as 12:15

A

B

C

19. John starts working in the garden at 5:30 PM and finishes 40 minutes later. What time does the clock show when John finishes his work? Represent this on a number line.

20. The clocks in the first column show different times. For each clock in the first column, select the correct answer.

	9:42	11:58	2:03
 Clock showing 11:58	○	○	○
 Clock showing 2:03	○	○	○
 Clock showing 9:42	○	○	○

Test ID	3M017
Test Name	Elapsed Time
Student Name	
Date	

1	Ⓐ Ⓑ Ⓒ Ⓓ	6	Ⓐ Ⓑ Ⓒ Ⓓ	11	Ⓐ Ⓑ Ⓒ Ⓓ	16	Answer in the space provided below.
2	Ⓐ Ⓑ Ⓒ Ⓓ	7	Ⓐ Ⓑ Ⓒ Ⓓ	12	Ⓐ Ⓑ Ⓒ Ⓓ	17	Answer in the space provided below.
3	Ⓐ Ⓑ Ⓒ Ⓓ	8	Ⓐ Ⓑ Ⓒ Ⓓ	13	Ⓐ Ⓑ Ⓒ Ⓓ	18	Ⓐ Ⓑ Ⓒ Ⓓ
4	Ⓐ Ⓑ Ⓒ Ⓓ	9	Ⓐ Ⓑ Ⓒ Ⓓ	14	Ⓐ Ⓑ Ⓒ Ⓓ	19	Answer in the space provided below.
5	Ⓐ Ⓑ Ⓒ Ⓓ	10	Ⓐ Ⓑ Ⓒ Ⓓ	15	Ⓐ Ⓑ Ⓒ Ⓓ	20	Ⓐ Ⓑ Ⓒ Ⓓ

16

	9:05	10:55	12:15
History 10:10	○	○	○
Math 8:20	○	○	○
Gym 11:30	○	○	○

17

Driving Start Time	Break Time
1:00	2:00
2:15	
3:30	4:30

19

Date of Completion:______________ Score:______________ 

Chapter 4 → Lesson 2: Elapsed Time

1. Cedric began reading his book at 9:12 AM. He finished at 10:02 AM. How long did it take him to read his book?

Ⓐ 50 minutes
Ⓑ 40 minutes
Ⓒ 48 minutes
Ⓓ 30 minutes

2. Samantha began eating her dinner at 7:11 PM and finished at 7:35 PM so that she could go to her room and play. How long did Samantha take to eat her dinner?

Ⓐ 34 minutes
Ⓑ 21 minutes
Ⓒ 24 minutes
Ⓓ 30 minutes

3. Tanya has after school tutoring from 3:00 PM until 3:25 PM. She began walking home at 3:31 PM and arrived at her house at 3:56 PM. How long did it take Tanya to walk home?

Ⓐ 31 minutes
Ⓑ 15 minutes
Ⓒ 56 minutes
Ⓓ 25 minutes

4. Doug loves to play video games. He started playing at 4:00 PM and did not finish until 5:27 PM. How long did Doug play video games?

Ⓐ 1 hour and 37 minutes
Ⓑ 1 hour and 27 minutes
Ⓒ 27 minutes
Ⓓ 2 hours and 27 minutes

5. Kelly has to clean her room before going to bed. She began cleaning her room at 6:12 PM. When she finished, it was 7:15 PM. How long did it take Kelly to clean her room?

Ⓐ 57 minutes
Ⓑ 53 minutes
Ⓒ 1 hour and 3 minutes
Ⓓ 1 hour and 15 minutes

6. Holly had a busy day. She attended a play from 7:06 PM until 8:13 PM. Then she went to dinner from 8:30 to 9:30 PM. How long did Holly attend the play?

   Ⓐ 57 minutes
   Ⓑ 2 hours and 27 minutes
   Ⓒ 46 minutes
   Ⓓ 1 hour and 7 minutes

7. Cara took her little brother to the park. They arrived at 3:11 PM and played until 4:37 PM. How long did Cara and her brother play at the park?

   Ⓐ 26 minutes
   Ⓑ 1 hour and 26 minutes
   Ⓒ 56 minutes
   Ⓓ 1 hour and 37 minutes

8. Arthur ran 5 miles. He began running at 8:19 AM and finished at 9:03 AM. How long did it take Arthur to run 5 miles?

   Ⓐ 44 minutes
   Ⓑ 45 minutes
   Ⓒ 40 minutes
   Ⓓ 54 minutes

9. Mr. Daniels wanted to see how fast he could wash the dishes. He began washing at 4:17 PM and finished at 4:32 PM. How long did it take Mr. Daniels to wash the dishes?

   Ⓐ 15 minutes
   Ⓑ 25 minutes
   Ⓒ 27 minutes
   Ⓓ 32 minutes

10. Sophia took a test that started at 3:28 PM. She finished the test at 4:11 PM. How long did it take Sophia to take her test?

    Ⓐ 37 minutes
    Ⓑ 47 minutes
    Ⓒ 33 minutes
    Ⓓ 43 minutes

11. Jonathan loves riding his bike, but he has to leave for football practice at 1:30 PM. If it is 1:11 PM now, how long does Jonathan have left to ride his bike before he has to leave for practice?

Ⓐ 9 minutes
Ⓑ 19 minutes
Ⓒ 21 minutes
Ⓓ 29 minutes

12. Mrs. Roberts loves to take a 20-minute nap on Saturdays. She was really tired when she went to sleep at 10:45 AM. She did not wake up until 11:25 AM. How long was Mrs. Roberts' long nap?

Ⓐ 40 minutes
Ⓑ 20 minutes
Ⓒ 60 minutes
Ⓓ 30 minutes

13. Spencer has to be at his piano lesson at noon. If it is now 11:29 AM, how long does Spencer have to get to his lesson?

Ⓐ 31 minutes
Ⓑ 1 minute
Ⓒ 29 minutes
Ⓓ 39 minutes

14. Look at the clocks below. How much time has elapsed between Clock A to Clock B?

Clock A

Clock B

Ⓐ 1 hour and 2 minutes
Ⓑ 1 hour and 12 minutes
Ⓒ 52 minutes
Ⓓ 42 minutes

15. Look at the clocks below. How much time has elapsed between Clock A to Clock B?

Clock A

Clock B

Ⓐ 52 minutes
Ⓑ 42 minutes
Ⓒ 12 minutes
Ⓓ 32 minutes

16. Tiana's daily schedule consists of classes that are 45 minutes long. The table shows what time some of her classes start. What time will each class end? Select the correct answer.

	9:05	10:55	12:15
History 10:10	◯	◯	◯
Math 8:20	◯	◯	◯
Gym 11:30	◯	◯	◯

17. Jonas is taking a long road trip. He drives for one hour and then stops and rests for 15 minutes. He repeats this until the end of the trip. Complete the table to show his schedule.

Driving Start Time	Break Time
1:00	2:00
2:15	
3:30	4:30

18. How much time has passed? Select all the correct answers.

Ⓐ 1 hour
Ⓑ 90 minutes
Ⓒ 2 hours
Ⓓ 120 minutes

19. Observe the two clocks. How many minutes have passed between the time shown in the first clock to the time in the second clock. Write your answer in the box given below.

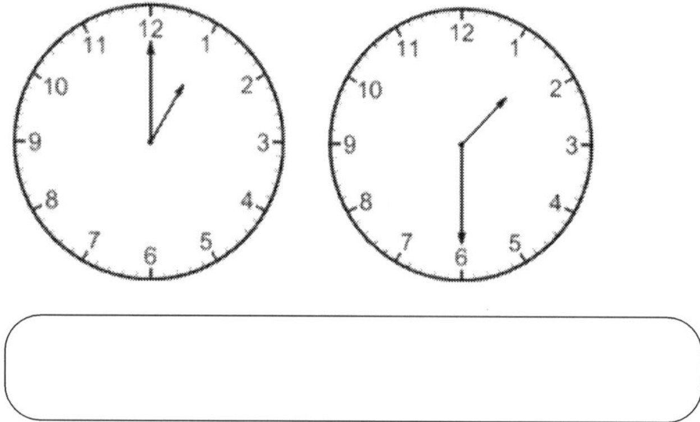

20. Tim went out to do some work. He left home at 11:30 AM and returned back at 3:45 PM. How long was he away from home? Circle the correct answer.

Ⓐ 3 hours and 15 minutes
Ⓑ 4 hours and 15 minutes
Ⓒ 3 hours and 45 minutes
Ⓓ 4 hours and 45 minutes

Test ID	3M018
Test Name	Liquid Volume & Mass
Student Name	
Date	

1	Ⓐ Ⓑ Ⓒ Ⓓ	6	Ⓐ Ⓑ Ⓒ Ⓓ	11	Ⓐ Ⓑ Ⓒ Ⓓ	16	Ⓐ Ⓑ Ⓒ Ⓓ
2	Ⓐ Ⓑ Ⓒ Ⓓ	7	Ⓐ Ⓑ Ⓒ Ⓓ	12	Ⓐ Ⓑ Ⓒ Ⓓ	17	Answer in the space provided below.
3	Ⓐ Ⓑ Ⓒ Ⓓ	8	Ⓐ Ⓑ Ⓒ Ⓓ	13	Ⓐ Ⓑ Ⓒ Ⓓ	18	Answer in the space provided below.
4	Ⓐ Ⓑ Ⓒ Ⓓ	9	Ⓐ Ⓑ Ⓒ Ⓓ	14	Ⓐ Ⓑ Ⓒ Ⓓ	19	Answer in the space provided below.
5	Ⓐ Ⓑ Ⓒ Ⓓ	10	Ⓐ Ⓑ Ⓒ Ⓓ	15	Ⓐ Ⓑ Ⓒ Ⓓ		

17

18

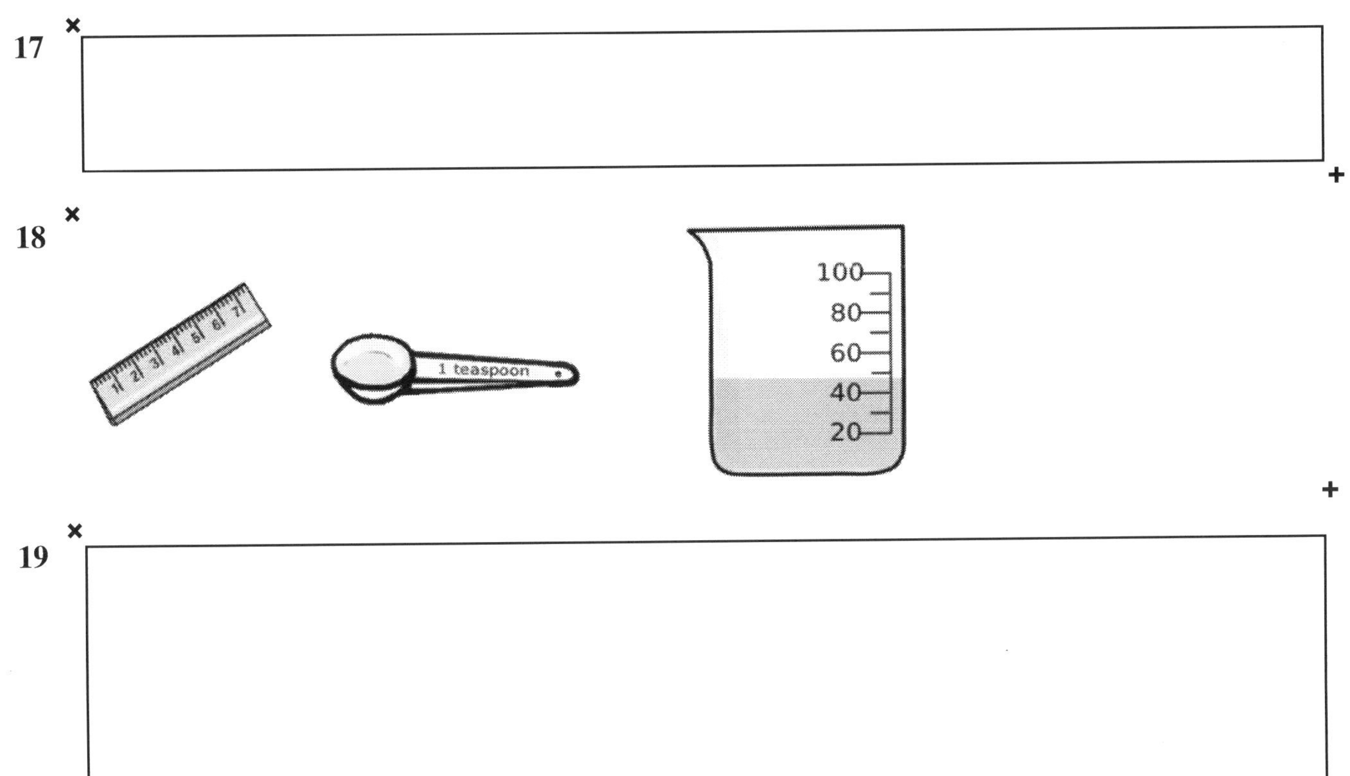

19

Date of Completion:______________ Score:______________ 

Chapter 4 → Lesson 3: Liquid Volume & Mass

1. "40 pounds" is printed at the bottom of a bag of sand. The number "40" is being used to __________ .

Ⓐ count
Ⓑ name
Ⓒ locate
Ⓓ measure

2. In the metric system, which is the best unit to measure the mass of a coffee table?

Ⓐ Milliliters
Ⓑ Kilograms
Ⓒ Grams
Ⓓ Liters

3. Which unit should be used to measure the amount of water in a small bowl?

Ⓐ Cups
Ⓑ Gallons
Ⓒ Inches
Ⓓ Tons

4. Which unit in the customary system is best suited to measure the weight of a coffee table?

Ⓐ Gallons
Ⓑ Pounds
Ⓒ Quarts
Ⓓ Ounces

5. Which of these units could be used to measure the capacity of a container?

Ⓐ pints
Ⓑ feet
Ⓒ pounds
Ⓓ millimeters

6. Which of these is a unit of mass?

Ⓐ liter
Ⓑ meter
Ⓒ gram
Ⓓ degree

7. Which of these units has the greatest capacity?

Ⓐ gallon
Ⓑ pint
Ⓒ cup
Ⓓ quart

8. Which of these might be the weight of an average sized 8 year-old child?

Ⓐ 15 pounds
Ⓑ 30 pounds
Ⓒ 65 pounds
Ⓓ 150 pounds

9. Volume is measured in __________ units.

Ⓐ cubic
Ⓑ liters
Ⓒ square
Ⓓ box

10. What is an appropriate unit to measure the weight of a dog?

Ⓐ tons
Ⓑ pounds
Ⓒ inches
Ⓓ gallons

11. What is an appropriate unit to measure the amount of water in a swimming pool?

Ⓐ teaspoons
Ⓑ cups
Ⓒ gallons
Ⓓ inches

12. What is an appropriate unit to measure the distance across a city?

Ⓐ centimeters
Ⓑ feet
Ⓒ inches
Ⓓ miles

13. What is an appropriate unit to measure the amount of salt in a cupcake recipe?

Ⓐ teaspoons
Ⓑ gallons
Ⓒ miles
Ⓓ kilograms

14. Which unit is the largest?

Ⓐ mile
Ⓑ centimeter
Ⓒ foot
Ⓓ inch

15. Which unit is the smallest?

Ⓐ kilometer
Ⓑ centimeter
Ⓒ millimeter
Ⓓ inch

16. Which units of measurement could be used to measure how much water a pot can hold? Select all the correct answers.

Ⓐ quarts
Ⓑ centimeter
Ⓒ liters
Ⓓ miles

17. Observe the figure given. How many cups of liquid does this measuring cup hold? Write your answer in the box given below.

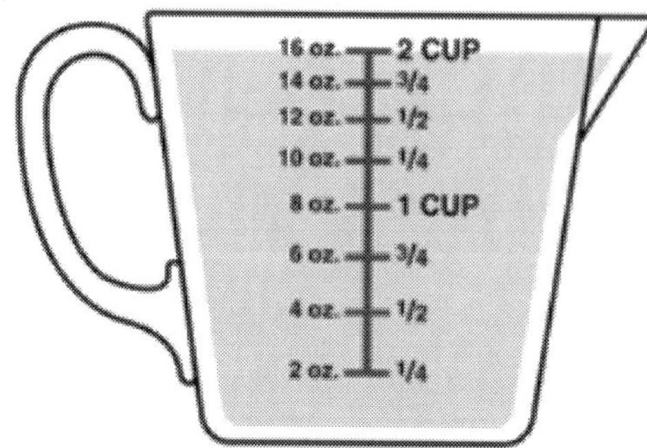

18. Circle the tool that should be used to measure a small amount of sugar.

19. There are 8 water coolers in a school. Each water cooler holds 7 liters of water. All the water coolers were filled up in the morning. In the evening 5 liters of water remained. How much water was consumed? Explain how you got the answer in the box below.

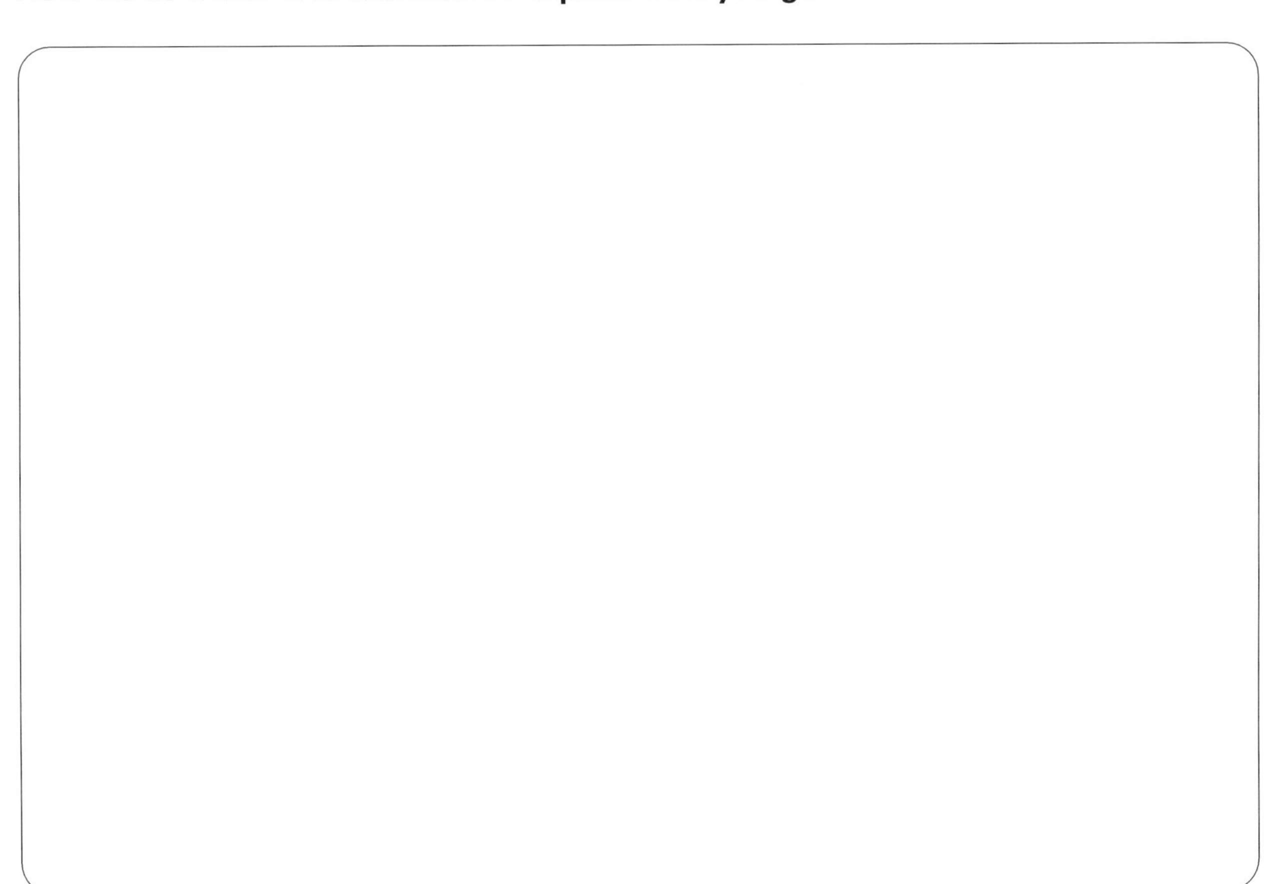

Test ID	3M019
Test Name	Graphs
Student Name	
Date	

1	Ⓐ Ⓑ Ⓒ Ⓓ	6	Ⓐ Ⓑ Ⓒ Ⓓ	11	Ⓐ Ⓑ Ⓒ Ⓓ	16	Answer in the space provided below.
2	Ⓐ Ⓑ Ⓒ Ⓓ	7	Ⓐ Ⓑ Ⓒ Ⓓ	12	Ⓐ Ⓑ Ⓒ Ⓓ	17	Answer in the space provided below.
3	Ⓐ Ⓑ Ⓒ Ⓓ	8	Ⓐ Ⓑ Ⓒ Ⓓ	13	Ⓐ Ⓑ Ⓒ Ⓓ	18	Ⓐ Ⓑ Ⓒ Ⓓ
4	Ⓐ Ⓑ Ⓒ Ⓓ	9	Ⓐ Ⓑ Ⓒ Ⓓ	14	Ⓐ Ⓑ Ⓒ Ⓓ	19	Answer in the space provided below.
5	Ⓐ Ⓑ Ⓒ Ⓓ	10	Ⓐ Ⓑ Ⓒ Ⓓ	15	Ⓐ Ⓑ Ⓒ Ⓓ	20	Answer in the space provided below.

16

	5	2	4
Total votes for Friday	○	○	○
Total votes for Wednesday	○	○	○
Total votes for Monday	○	○	○

17

Type of Bat	Number of Bats Needed
Wood	
Aluminum	
Plastic	

19

Coins in John's Piggy Bank		
Coin	Tally	Number of Coins
Penny	卌 卌 卌 卌 II	
Nickel	卌 卌 卌 III	
Dime	卌 卌 IIII	
Quarter	卌 卌 卌 I	

20

Date of Completion:______________ Score:______________

Chapter 4 → Lesson 4: Graphs

1.

Class Survey
Should there be a field trip?

	Yes	No
Mr. A's class	𝍸 𝍸 ||||	𝍸 ||
Mrs. B's class	𝍸 𝍸 𝍸	𝍸 𝍸 |||
Mr. C's class	𝍸 𝍸 |	𝍸 𝍸 |
Mrs. D's class	𝍸 𝍸 ||	𝍸 𝍸

Four 3rd grade classes in Hill Elementary School were surveyed to find out if they wanted to go on a field trip at the end of the school year. The tally table above was used to record the votes.
How many kids voted "Yes" in Mrs. B's class?

Ⓐ **28 kids**
Ⓑ **15 kids**
Ⓒ **13 kids**
Ⓓ **23 kids**

2.

Should there be a field trip?

	Yes	No
Mr. A's class	14	7
Mrs. B's class	15	13
Mr. C's class	11	11
Mrs. D's class	12	10
Total	52	41

Four 3rd grade classes in Hill Elementary School were surveyed to find out if they wanted to go on a field trip at the end of the school year. The table above shows the results of the survey.
How many kids voted "Yes" in Mr. A's class?

Ⓐ **7 kids**
Ⓑ **15 kids**
Ⓒ **14 kids**
Ⓓ **21 kids**

3.

Should there be a field trip?		
	Yes	No
Mr. A's class	14	7
Mrs. B's class	15	13
Mr. C's class	11	11
Mrs. D's class	12	10
Total	52	41

Four 3rd grade classes in Hill Elementary School were surveyed to find out if they wanted to go on a field trip at the end of the school year. The table above shows the results of the survey.
How many kids altogether voted "No" for the field trip?

Ⓐ **82 kids**
Ⓑ **11 kids**
Ⓒ **52 kids**
Ⓓ **41 kids**

4. **The students in Mr. Donovan's class were surveyed to find out their favorite school subjects. The results are shown in the pictograph. Use the pictograph to answer the following question:**
How many students chose either science or math?

Our Favorite Subjects

Math	OOOO
Reading	OO
Science	OOO
History	O
Other	OO

Key: O = 2 votes

Ⓐ **6 students**
Ⓑ **7 students**
Ⓒ **14 students**
Ⓓ **2 students**

5.

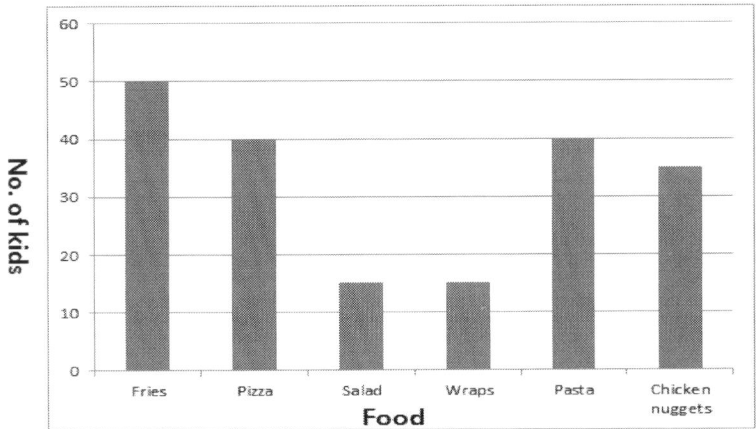

The third graders in Valley Elementary School were asked to pick their favorite food from 6 choices. The results are shown in the bar graph.
Which food was the favorite of the most children?

Ⓐ Pizza
Ⓑ Pasta
Ⓒ Fries
Ⓓ Salad

6.

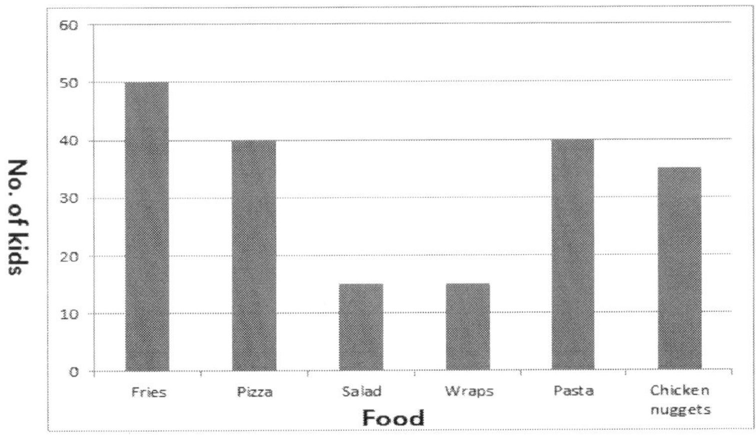

The third graders in Valley Elementary School were asked to pick their favorite food from 6 choices. The results are shown in the bar graph.

What are the 2 foods that kids like the least?

Ⓐ Fries and Pizza
Ⓑ Pizza and Pasta
Ⓒ Pasta and Chicken Nuggets
Ⓓ Salad and Wraps

7.

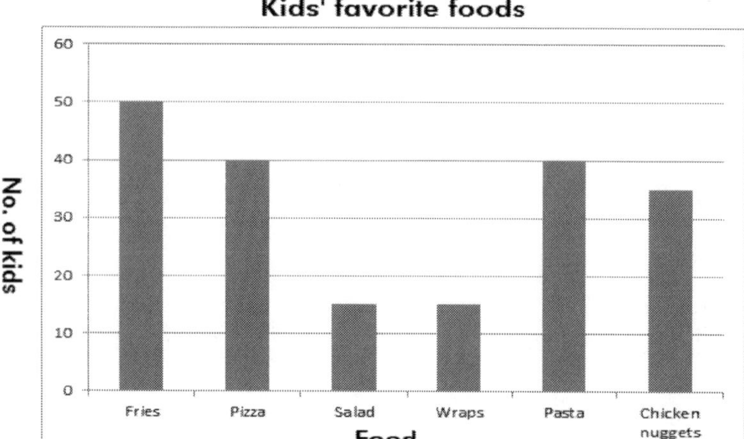

The third graders in Valley Elementary School were asked to pick their favorite food from 6 choices. The results are shown in the bar graph.
How many kids chose pasta?

Ⓐ 50 kids
Ⓑ 15 kids
Ⓒ 40 kids
Ⓓ 35 kids

8.

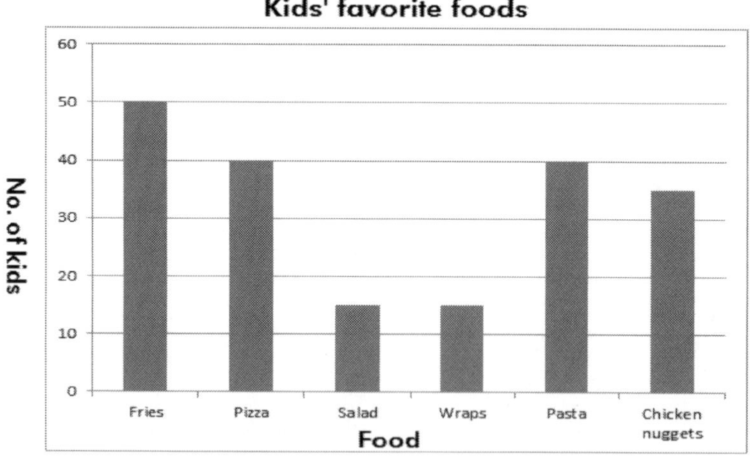

The third graders in Valley Elementary School were asked to pick their favorite food from 6 choices. The results are shown in the bar graph.

How many more kids prefer fries than pizza?

Ⓐ 50 more kids
Ⓑ 10 more kids
Ⓒ 1 more kid
Ⓓ 15 more kids

9. 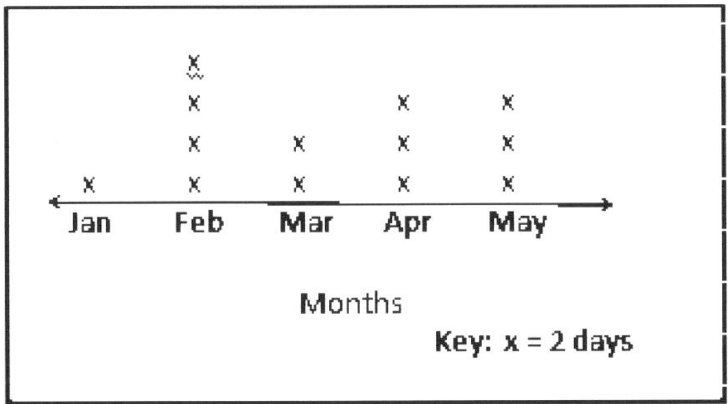

The line plot shows the number of days it rained in New Jersey from January through May. What should be the title of the above graph?

Ⓐ Line plot
Ⓑ Rainy Days in New Jersey
Ⓒ Months
Ⓓ 2 days

10. Rainy Days in New Jersey

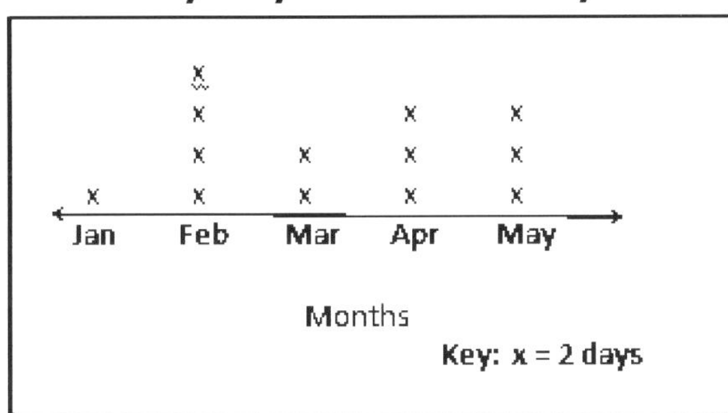

Which of the following statements is TRUE for the above graph?

Ⓐ The graph shows the total liters of water collected in New Jersey from January to May.
Ⓑ The graph shows rainfall for USA from January to May.
Ⓒ The graph shows the average temperature during the 5 month period.
Ⓓ The graph shows the number of rainy days in New Jersey from January to May.

11. 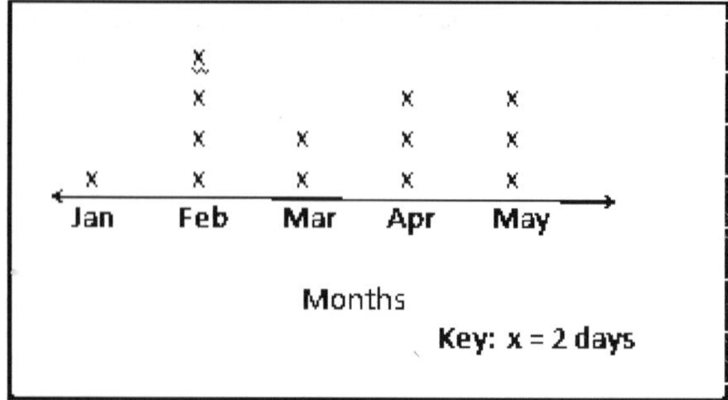

According to the graph, which month had the most rainy days?

Ⓐ **March**
Ⓑ **February**
Ⓒ **January**
Ⓓ **April**

12. **A survey was taken to find out the favorite sports of third graders in a particular class. The results are shown in the tally table. Use the table to answer the following question: How many students were surveyed altogether?**

Our Favorite Sports

Soccer	𝍸 |
Tennis	||||
Baseball	𝍸 |||
Hockey	||||
Other	|||

Ⓐ **20 students**
Ⓑ **25 students**
Ⓒ **24 students**
Ⓓ **27 students**

13. A survey was taken to find out the favorite sports of third graders in a particular class. The results are shown in the tally table. Use the table to answer the following question: How many more students chose soccer than chose hockey?

Our Favorite Sports

Soccer	𝍸 |
Tennis	||||
Baseball	𝍸 |||
Hockey	||||
Other	|||

Ⓐ 6 students
Ⓑ 4 students
Ⓒ 2 students
Ⓓ 3 students

14. A survey was taken to find out the favorite sports of third graders in a particular class. The results are shown in the tally table. Use the table to answer the following question: How many students chose baseball as their favorite sport?

Our Favorite Sports

Soccer	𝍸 |
Tennis	||||
Baseball	𝍸 |||
Hockey	||||
Other	|||

Ⓐ 9 students
Ⓑ 8 students
Ⓒ 3 students
Ⓓ 6 students

15. A survey was taken to find out the favorite sports of third graders in a particular class. The results are shown in the tally table. Use the table to answer the following question: Which two sports were chosen by the same number of students?

Our Favorite Sports

Soccer	~~||||~~ |
Tennis	||||
Baseball	~~||||~~ |||
Hockey	||||
Other	|||

Ⓐ soccer and tennis
Ⓑ soccer and baseball
Ⓒ hockey and soccer
Ⓓ hockey and tennis

16. Mrs. Brown's class voted on which day they will have a class party. Look at the graph. Each figure represents 1 student. Match the correct answers to the number of votes.

	5	2	4
Total votes for Friday	◯	◯	◯
Total votes for Wednesday	◯	◯	◯
Total votes for Monday	◯	◯	◯

17. Coach Dennis is creating a graph. He wants to purchase bats for his team. He needs to purchase 8 bats. He needs half of the bats to be made of wood. The other half will be made of aluminum. He decides that he will purchase 3 more bats made of plastic as well. Complete the table by filling in the correct answers.

Type of Bat	Number of Bats Needed
Wood	
Aluminum	
Plastic	

18. Vicki is planting a flower garden. The graph above shows the number of flowers to be planted in the garden. Which of the following statements are true? Select all the correct answers.

= 2 Hibiscus flowers

= 2 Rose flowers

= 2 Iris flowers

Ⓐ The count of Rose flowers is 14
Ⓑ The Count of Rose flowers is 7
Ⓒ The count of Hibiscus flowers is 4
Ⓓ The count of Iris flowers is 6

19. **Find the total number of each coin. Use the tally chart to draw a bar graph.**

Coins in John's Piggy Bank		
Coin	**Tally**	**Number of Coins**
Penny	卌 卌 卌 卌 ||	
Nickel	卌 卌 卌 |||	
Dime	卌 卌 ||||	
Quarter	卌 卌 卌 |	

20. From the Venn Diagram given below, represent the number of people who only own cats as pet to the number of people who own only dogs as a pet in the form of a fraction $\frac{a}{b}$ (ratio of number of people owning cats to dogs)

Test ID	3M020
Test Name	Measuring Length
Student Name	
Date	

1	Ⓐ Ⓑ Ⓒ Ⓓ	6	Ⓐ Ⓑ Ⓒ Ⓓ	11	Ⓐ Ⓑ Ⓒ Ⓓ	16	Ⓐ Ⓑ Ⓒ Ⓓ
2	Ⓐ Ⓑ Ⓒ Ⓓ	7	Ⓐ Ⓑ Ⓒ Ⓓ	12	Ⓐ Ⓑ Ⓒ Ⓓ	17	Answer in the space provided below.
3	Ⓐ Ⓑ Ⓒ Ⓓ	8	Ⓐ Ⓑ Ⓒ Ⓓ	13	Ⓐ Ⓑ Ⓒ Ⓓ	18	Answer in the space provided below.
4	Ⓐ Ⓑ Ⓒ Ⓓ	9	Ⓐ Ⓑ Ⓒ Ⓓ	14	Ⓐ Ⓑ Ⓒ Ⓓ	19	Answer in the space provided below.
5	Ⓐ Ⓑ Ⓒ Ⓓ	10	Ⓐ Ⓑ Ⓒ Ⓓ	15	Ⓐ Ⓑ Ⓒ Ⓓ		

17

18

Measurement in inches	Measurement in half inches	Measurement in quarter inches
$3\frac{1}{2}$ inches	7 half inches	14 quarter inches
$2\frac{1}{2}$ inches		
	11 half inches	
		26 quarter inches

19

	4	6	8
How many fish are $16\frac{1}{2}$ inches long?	○	○	○
How many more fish are 16 inches long than 17 inches?	○	○	○
How many fish are less than $15\frac{3}{4}$ inches long?	○	○	○

Date of Completion:______________ Score:______________

Chapter 4 → Lesson 5: Measuring Length

1. Which of these units is part of the metric system?

   Ⓐ Foot
   Ⓑ Mile
   Ⓒ Kilometer
   Ⓓ Yard

2. Which metric unit is closest in length to one yard?

   Ⓐ decimeter
   Ⓑ meter
   Ⓒ millimeter
   Ⓓ kilometer

3. Which of these is the best estimate for the length of a table?

   Ⓐ 2 decimeters
   Ⓑ 2 centimeters
   Ⓒ 2 meters
   Ⓓ 2 kilometers

4. What unit should you use to measure the length of a book?

   Ⓐ Kilometers
   Ⓑ Meters
   Ⓒ Centimeters
   Ⓓ Grams

5. About how long is a new pencil?

   Ⓐ 8 inches
   Ⓑ 8 feet
   Ⓒ 8 yards
   Ⓓ 8 miles

6. Which of these is the best estimate for the length of a football?

   Ⓐ 1 foot
   Ⓑ 2 feet
   Ⓒ 6 feet
   Ⓓ 4 feet

7. Complete the following statement.
   The length of a football field is __________.

   Ⓐ less than one meter
   Ⓑ greater than one meter
   Ⓒ about one meter
   Ⓓ impossible to measure

8. Complete the following statement.
   An adult's pointer finger is about one ________ wide.

   Ⓐ meter
   Ⓑ kilometer
   Ⓒ millimeter
   Ⓓ centimeter

9. Complete the following statement.
   The distance between two cities would most likely be measured in ____________.

   Ⓐ hours
   Ⓑ miles
   Ⓒ yards
   Ⓓ square inches

10. A ribbon is 25 centimeters long. About how many inches long is it?

   Ⓐ 2
   Ⓑ 25
   Ⓒ 10
   Ⓓ 50

11.

   How long is this object?

   Ⓐ 4 inches
   Ⓑ 8 inches
   Ⓒ 10 inches
   Ⓓ 12 inches

12.

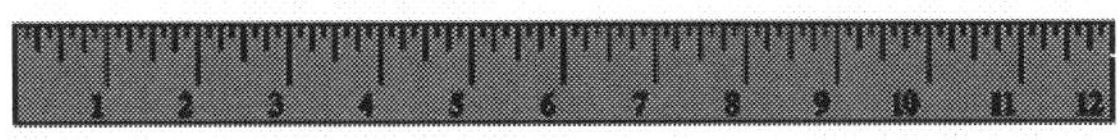

How long is this object?

Ⓐ 2 inches
Ⓑ 5 inches
Ⓒ 3 inches
Ⓓ 1 inch

13.

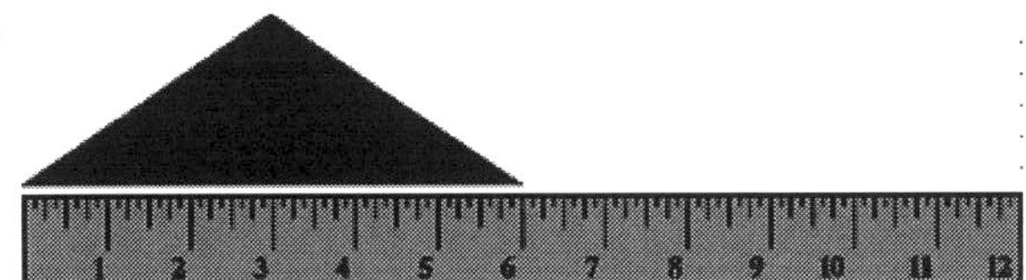

How long is this object?

Ⓐ 6 inches
Ⓑ 5 and a half inches
Ⓒ 6 and a half inches
Ⓓ 7 inches

14. Which statement is correct?

Ⓐ 1 inch > 1 mile
Ⓑ 1 inch > 1 centimeter
Ⓒ 1 foot < 1 inch
Ⓓ 1 mile < 1 foot

15. Which is correct?

Ⓐ 12 inches > 1 foot
Ⓑ 12 inches < 1 foot
Ⓒ 12 inches = 1 foot
Ⓓ 9 inches = 1 foot

16. Which of the following could be measured with a ruler? Select all correct answers.

Ⓐ water in a bowl
Ⓑ a football field
Ⓒ a carrot
Ⓓ a crayon

17. **Observe the figure. How long is the pencil when measured in inches?**

18. **Fill in the correct answer in the blanks shown in the table.**

Measurement in inches	Measurement in half inches	Measurement in quarter inches
$3\frac{1}{2}$ inches	7 half inches	14 quarter inches
$2\frac{1}{2}$ inches		
	11 half inches	
		26 quarter inches

19. **Use the line plot to answer the questions given in the first column.**

Lengths of Fish

Instruction:
X = 2 fishes

	4	6	8
How many fish are $16\frac{1}{2}$ inches long?	○	○	○
How many more fish are 16 inches long than 17 inches?	○	○	○
How many fish are less than $15\frac{3}{4}$ inches long?	○	○	○

Test ID	3M021
Test Name	Area
Student Name	
Date	

1	(A) (B) (C) (D)	6	(A) (B) (C) (D)	11	Answer in the space provided below.
2	(A) (B) (C) (D)	7	(A) (B) (C) (D)	12	Answer in the space provided below.
3	(A) (B) (C) (D)	8	(A) (B) (C) (D)	13	Answer in the space provided below.
4	(A) (B) (C) (D)	9	(A) (B) (C) (D)	14	(A) (B) (C) (D)
5	(A) (B) (C) (D)	10	(A) (B) (C) (D)	15	(A) (B) (C) (D)

11

	Yes	No
Window panes		
A ball		
Bathroom tile		
A banana		

12

Figure	Area
Figure A	
Figure B	
Figure C	

13

Chapter 4 → Lesson 6: Area

1. The area of a plane figure is measured in __________ units.

Ⓐ cubic
Ⓑ meter
Ⓒ square
Ⓓ box

2. Which of these objects has an area of about 1 square inch?

Ⓐ a sheet of writing paper
Ⓑ a beach towel
Ⓒ a dollar bill
Ⓓ a postage stamp

3. Mr. Parker wants to cover a mural with cloth. The mural is 12 inches long and 20 inches wide. How many square inches of cloth does Mr. Parker need?

Ⓐ 240 square inches
Ⓑ 32 square inches
Ⓒ 120 square inches
Ⓓ 220 square inches

4.

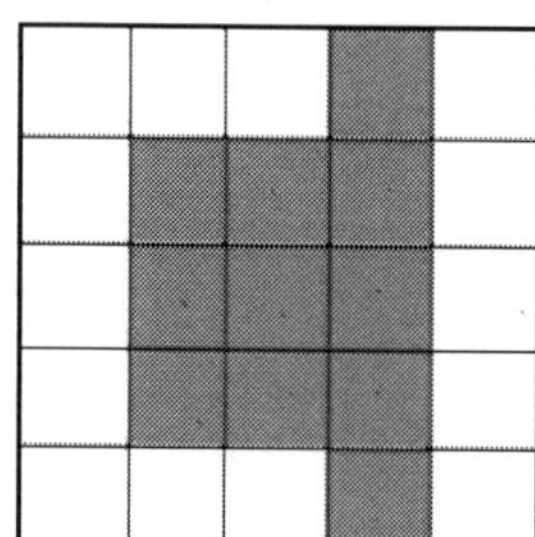

=1 Square Unit

What is the area of the shaded region?

Ⓐ 10 square units
Ⓑ 8 square units
Ⓒ 11 square units
Ⓓ 15 square units

5. Find the area of this figure.

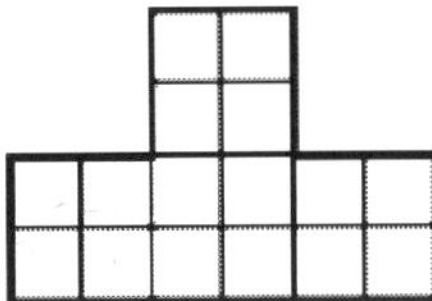

□ =1 Square Unit

Ⓐ 22 square units
Ⓑ 20 square units
Ⓒ 18 square units
Ⓓ 16 square units

6. Find the area of this figure.

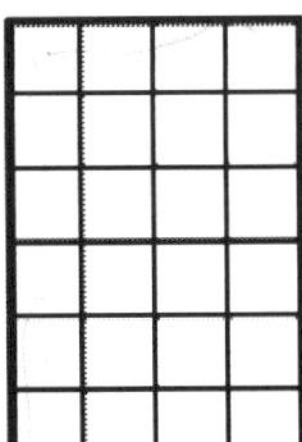

□ =1 Square Unit

Ⓐ 22 square units
Ⓑ 20 square units
Ⓒ 24 square units
Ⓓ 28 square units

7. Find the area of the shaded region.

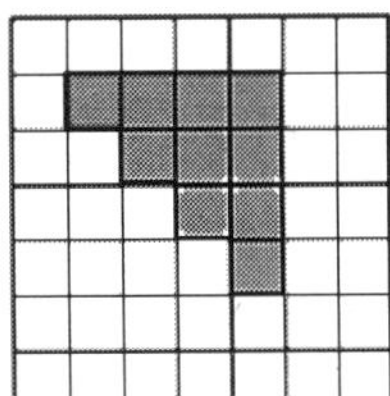

□ =1 Square Unit

Ⓐ 11 square units
Ⓑ 10 square units
Ⓒ 16 square units
Ⓓ 9 square units

8. Find the area of the shaded region.

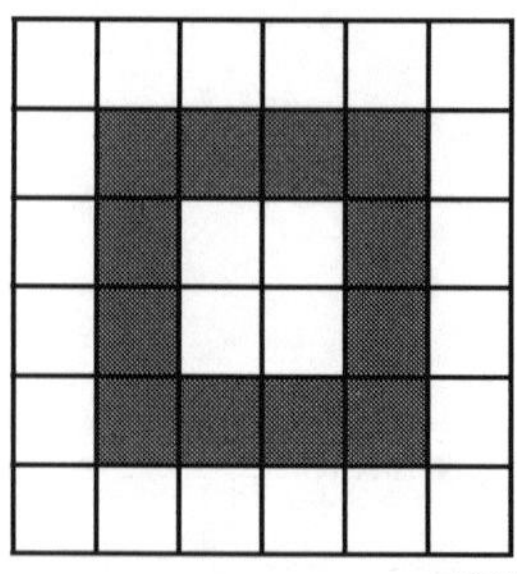

= 1 Square Unit

Ⓐ 16 square units
Ⓑ 12 square units
Ⓒ 10 square units
Ⓓ 11 square units

9. Find the area of the shaded region.

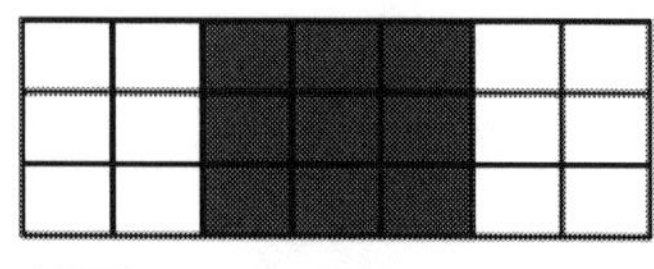

Key: = 1 Square Unit

Ⓐ 5 square units
Ⓑ 6 square units
Ⓒ 8 square units
Ⓓ 9 square units

10. The area of Karen's rectangular room is 72 sq. ft. If the length of the room is 8 ft. What is its width?

Ⓐ 8 ft.
Ⓑ 6 ft.
Ⓒ 7 ft.
Ⓓ 9 ft.

11. Can these items be measured in square units? Select yes or no.

	Yes	No
Window panes		
A ball		
Bathroom tile		
A banana		

12. Find the area of the shaded region in each figure. Each box is 1 square unit. Write your answers in the blank boxes in the table.

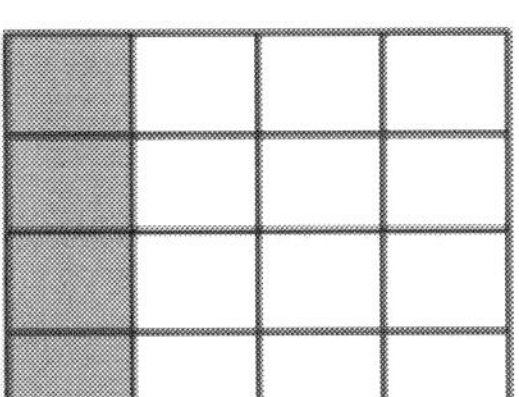

Figure A

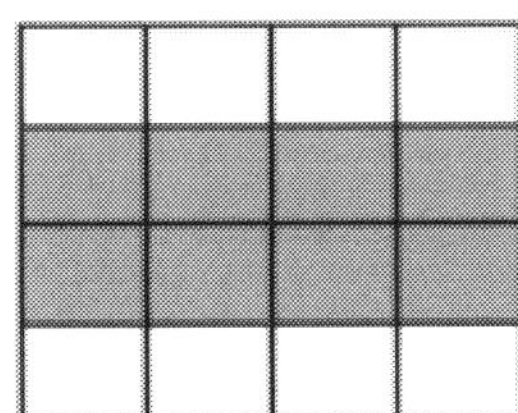

Figure B

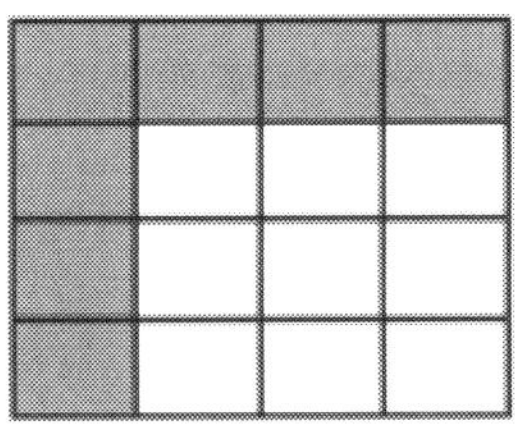

Figure C

Figure	Area
Figure A	
Figure B	
Figure C	

13. Find the area of the figure. Write your answer in the box given below.

6 feet

9 feet

square feet

14. Which of the following are possible ways to find the area of this figure? Each box is 1 square unit. Select all correct answers.

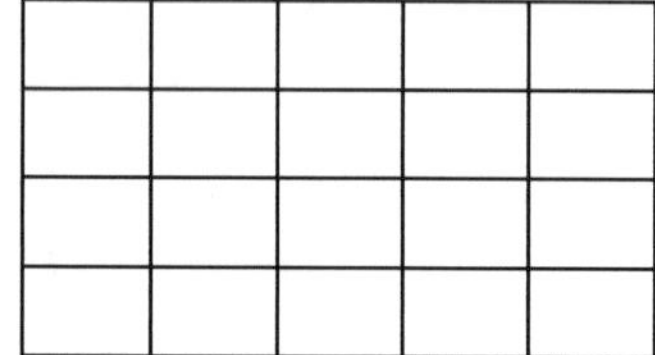

Ⓐ Count the total number of square units
Ⓑ Multiplying the length by the width of the figure
Ⓒ Multiplying the number of square units by 2
Ⓓ Subtracting the length of the figure from the width

15. The area of a rectangle A is 75 sq. cm. The area of square B is one third the area of the rectangle A. What is the side length of the square B? Circle the correct answer.

Ⓐ 7 cm
Ⓑ 5 cm
Ⓒ 4 cm
Ⓓ 6 cm

Test ID	3M022
Test Name	Relating Area to Addition & Multiplication
Student Name	
Date	

1	Ⓐ Ⓑ Ⓒ Ⓓ	6	Ⓐ Ⓑ Ⓒ Ⓓ	11	Ⓐ Ⓑ Ⓒ Ⓓ
2	Ⓐ Ⓑ Ⓒ Ⓓ	7	Ⓐ Ⓑ Ⓒ Ⓓ	12	Ⓐ Ⓑ Ⓒ Ⓓ
3	Ⓐ Ⓑ Ⓒ Ⓓ	8	Ⓐ Ⓑ Ⓒ Ⓓ	13	Ⓐ Ⓑ Ⓒ Ⓓ
4	Ⓐ Ⓑ Ⓒ Ⓓ	9	Ⓐ Ⓑ Ⓒ Ⓓ	14	Ⓐ Ⓑ Ⓒ Ⓓ
5	Ⓐ Ⓑ Ⓒ Ⓓ	10	Ⓐ Ⓑ Ⓒ Ⓓ	15	Ⓐ Ⓑ Ⓒ Ⓓ

Chapter 4 → Lesson 7: Relating Area to Addition & Multiplication

1. Find the area of the rectangle below.

3 feet

29 feet

Ⓐ 87 square feet
Ⓑ 32 square feet
Ⓒ 64 square feet
Ⓓ 128 square feet

2. Find the area of the rectangle below.

12 yards

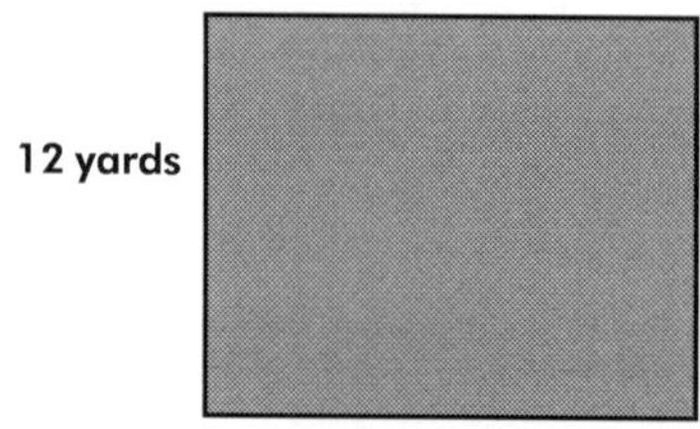

15 yards

Ⓐ 108 square yards
Ⓑ 54 square yards
Ⓒ 27 square yards
Ⓓ 180 square yards

3. How could the area of this figure be calculated?

33 inches

63 inches

Ⓐ Multiply 63 x 33 x 63 x 33
Ⓑ Add 63 + 33 + 63 + 33
Ⓒ Multiply 63 x 33
Ⓓ Multiply 2 x 63 x 33

4. Find the area of the rectangle below.

5 meters

10 meters

Ⓐ 75 square meters
Ⓑ 50 square meters
Ⓒ 15 square meters
Ⓓ 30 square meters

5. Find the area of the rectangle below.

16 yards
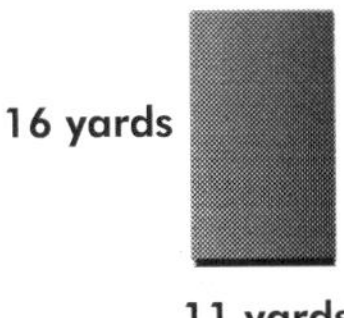
11 yards

Ⓐ 176 square yards
Ⓑ 27 square yards
Ⓒ 54 square yards
Ⓓ 2,916 square yards

6. Find the area of the rectangle below.

3 inches

2 inches + 3 inches

Ⓐ 18 square inches
Ⓑ 15 square inches
Ⓒ 9 square inches
Ⓓ 6 square inches

7. Find the area of the rectangle below.

7 feet

2 feet + 1 foot

Ⓐ 16 square feet
Ⓑ 14 square feet
Ⓒ 9 square feet
Ⓓ 21 square feet

8. **Find the area of the rectangle below.**

Ⓐ **84 square meters**
Ⓑ **48 square meters**
Ⓒ **36 square meters**
Ⓓ **72 square meters**

9. **Find the area of the rectangle below.**

Ⓐ **12 square inches**
Ⓑ **10 square inches**
Ⓒ **25 square inches**
Ⓓ **35 square inches**

10. **Find the area of the rectangle below.**

Ⓐ **84 square yards**
Ⓑ **182 square yards**
Ⓒ **26 square yards**
Ⓓ **19 square yards**

11. **The city wants to plant grass in a park. The park is 20 feet by 50 feet. How much grass will they need to cover the entire park?**

Ⓐ **100 square feet**
Ⓑ **500 square feet**
Ⓒ **1,000 square feet**
Ⓓ **1,100 square feet**

12. Brenda wants to purchase a rug for her room. Her room is a rectangle that measures 7 yards by 6 yards. What is the area of her room?

Ⓐ 42 square yards
Ⓑ 48 square yards
Ⓒ 36 square yards
Ⓓ 26 square yards

13. Joan wants to cover her backyard with flowers. If her backyard is 30 feet long and 20 feet wide, what is the area that needs to be covered in flowers?

Ⓐ 60 square feet
Ⓑ 500 square feet
Ⓒ 600 square feet
Ⓓ 100 square feet

14. Bethany decided to paint the four walls in her room. If each wall measures 20 feet by 10 feet, how many total square feet will she need to paint?

Ⓐ 400 square feet
Ⓑ 200 square feet
Ⓒ 800 square feet
Ⓓ 600 square feet

15. Seth wants to cover his table top with a piece of fabric. His table is 2 meters long and 4 meters wide. How much fabric does Seth need?

Ⓐ 6 square meters
Ⓑ 10 square meters
Ⓒ 8 square meters
Ⓓ 16 square meters

Test ID	3M023
Test Name	Perimeter
Student Name	
Date	

1	Ⓐ Ⓑ Ⓒ Ⓓ	11	Ⓐ Ⓑ Ⓒ Ⓓ	21	Ⓐ Ⓑ Ⓒ Ⓓ
2	Ⓐ Ⓑ Ⓒ Ⓓ	12	Ⓐ Ⓑ Ⓒ Ⓓ	22	Ⓐ Ⓑ Ⓒ Ⓓ
3	Ⓐ Ⓑ Ⓒ Ⓓ	13	Ⓐ Ⓑ Ⓒ Ⓓ	23	Ⓐ Ⓑ Ⓒ Ⓓ
4	Ⓐ Ⓑ Ⓒ Ⓓ	14	Ⓐ Ⓑ Ⓒ Ⓓ	24	Ⓐ Ⓑ Ⓒ Ⓓ
5	Ⓐ Ⓑ Ⓒ Ⓓ	15	Ⓐ Ⓑ Ⓒ Ⓓ	25	Ⓐ Ⓑ Ⓒ Ⓓ
6	Ⓐ Ⓑ Ⓒ Ⓓ	16	Ⓐ Ⓑ Ⓒ Ⓓ	26	Ⓐ Ⓑ Ⓒ Ⓓ
7	Ⓐ Ⓑ Ⓒ Ⓓ	17	Ⓐ Ⓑ Ⓒ Ⓓ	27	Answer in the space provided below.
8	Ⓐ Ⓑ Ⓒ Ⓓ	18	Ⓐ Ⓑ Ⓒ Ⓓ	28	Answer in the space provided below.
9	Ⓐ Ⓑ Ⓒ Ⓓ	19	Ⓐ Ⓑ Ⓒ Ⓓ	29	Answer in the space provided below.
10	Ⓐ Ⓑ Ⓒ Ⓓ	20	Ⓐ Ⓑ Ⓒ Ⓓ	30	Answer in the space provided below.

27

28

2 cm 5 cm 4 cm

29

30

	15 sq. cm.	10 sq. cm.	12 sq. cm.
Perimeter = 16 cm	○	○	○
Perimeter = 14 cm	○	○	○
Perimeter = 22 cm	○	○	○

Date of Completion:_______________ Score:_______________

Chapter 4 → Lesson 8: Perimeter

1. What is meant by the "perimeter" of a shape?

   Ⓐ The distance from the center of a plane figure to its edge
   Ⓑ The distance from one corner of a plane figure to an opposite corner
   Ⓒ The distance around the outside of a plane figure
   Ⓓ The amount of space covered by a plane figure

2. Complete the following statement.
   Two measurements associated with plane figures are _______________.

   Ⓐ perimeter and volume
   Ⓑ perimeter and area
   Ⓒ volume and area
   Ⓓ weight and volume

3.

   This rectangle is 4 units long and one unit wide. What is its perimeter?

   Ⓐ 10 units
   Ⓑ 4 units
   Ⓒ 5 units
   Ⓓ 8 units

4.

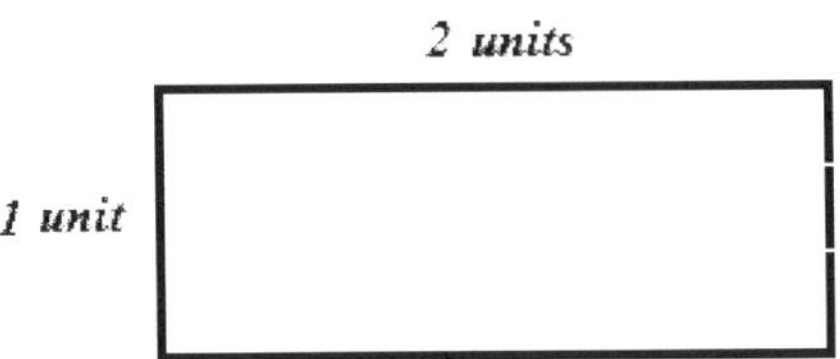

   What is the perimeter of the rectangle?

   Ⓐ 5 units
   Ⓑ 6 units
   Ⓒ 3 units
   Ⓓ 2 units

5. 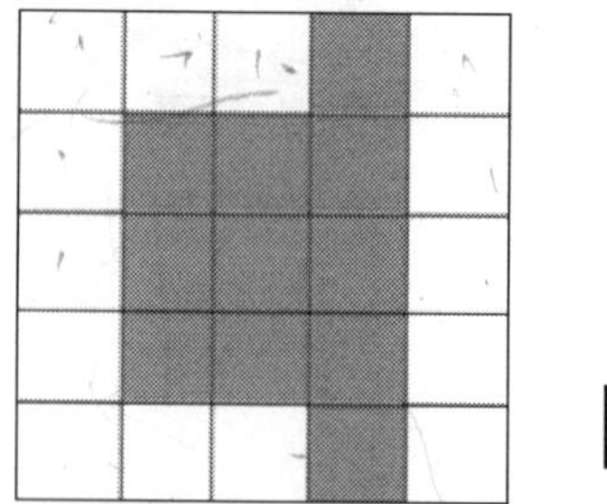

What is the perimeter of the shaded region in the above figure?

Ⓐ 16 units
Ⓑ 15 units
Ⓒ 11 units
Ⓓ 10 units

6. The perimeter of this rhombus is 20 units. How long is each of its sides?

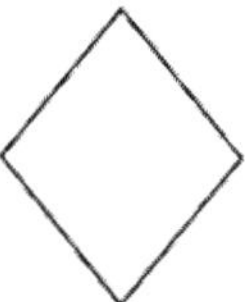

Ⓐ 4 units
Ⓑ 10 units
Ⓒ 5 units
Ⓓ This cannot be determined.

7. Each side of this rhombus measures 3 centimeters. What is its perimeter?

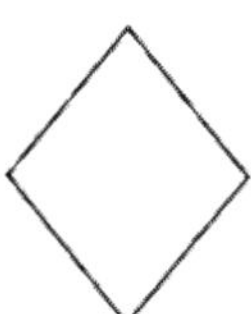

Ⓐ 3 centimeters
Ⓑ 12 centimeters
Ⓒ 9 centimeters
Ⓓ 6 centimeters

8. This square has a perimeter of 80 units. How long is each of its sides?

Ⓐ 8 units
Ⓑ 10 units
Ⓒ 20 units
Ⓓ 40 units

9. Find the perimeter of this figure.

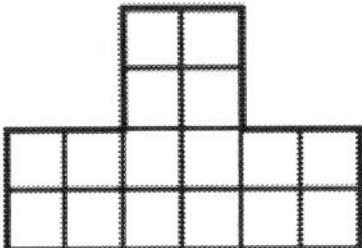

= 1 Square Unit

Ⓐ 20 units
Ⓑ 18 units
Ⓒ 16 units
Ⓓ 22 units

10. Joan wants to cover the outside border of her backyard with flowers. If her backyard is 30 feet long and 15 feet wide, how many feet of flowers does she need to plant?

Ⓐ 450 feet
Ⓑ 90 feet
Ⓒ 60 feet
Ⓓ 30 feet

11. Find the perimeter of this figure.

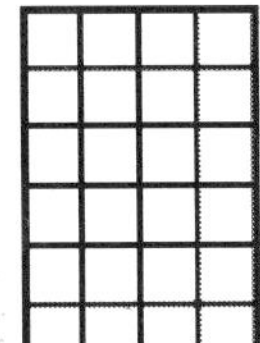

= 1 Square Unit

Ⓐ 20 units
Ⓑ 18 units
Ⓒ 24 units
Ⓓ 22 units

12. Brenda wants to place rope around a large field in order to play a game. The field is a rectangle that measures 23 yards by 32 yards. How much rope does Brenda need?

Ⓐ 64 yards
Ⓑ 736 yards
Ⓒ 110 yards
Ⓓ 55 yards

13. Find the perimeter of the following rectangle.

Ⓐ 54 feet
Ⓑ 27 feet
Ⓒ 58 feet
Ⓓ 180 feet

14. Find the perimeter of the following rectangle.

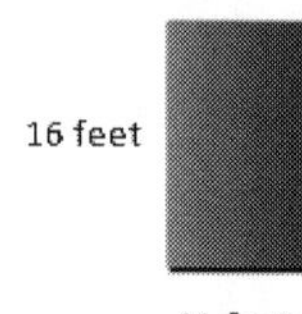

11 feet

Ⓐ 42 feet
Ⓑ 176 feet
Ⓒ 27 feet
Ⓓ 54 feet

15. Find the perimeter of the shaded region.

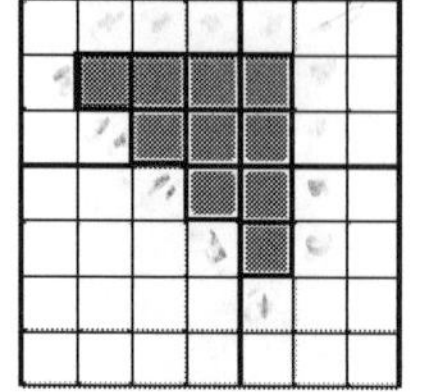

☐ =1 Square Unit

Ⓐ 10 units
Ⓑ 13 units
Ⓒ 15 units
Ⓓ 16 units

16. Find the perimeter of the following rectangle.

Ⓐ 32 feet
Ⓑ 87 feet
Ⓒ 172 feet
Ⓓ 64 feet

17. The city is building a fence around a park. The park is 20 feet by 50 feet. How many feet of fencing do they need?

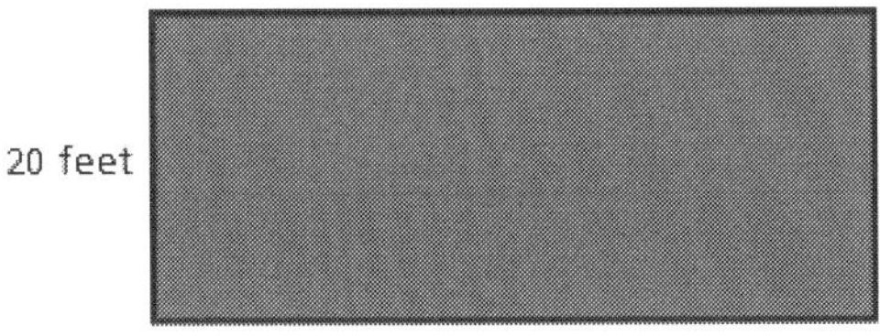

Ⓐ 100 feet
Ⓑ 120 feet
Ⓒ 140 feet
Ⓓ 70 feet

18. Find the perimeter of the following rectangle.

Ⓐ 86 inches
Ⓑ 172 inches
Ⓒ 1,749 inches
Ⓓ 50 inches

19. **Find the perimeter of the following rectangle.**

10 meters

Ⓐ **15 meters**
Ⓑ **25 meters**
Ⓒ **30 meters**
Ⓓ **50 meters**

20. **The perimeter of the following object is 16 feet. Find the length of the missing side.**

x

Ⓐ **x = 5 feet**
Ⓑ **x = 10 feet**
Ⓒ **x = 13 feet**
Ⓓ **x = 6 feet**

21. **The perimeter of the following object is 20 feet. Find the length of the missing side.**

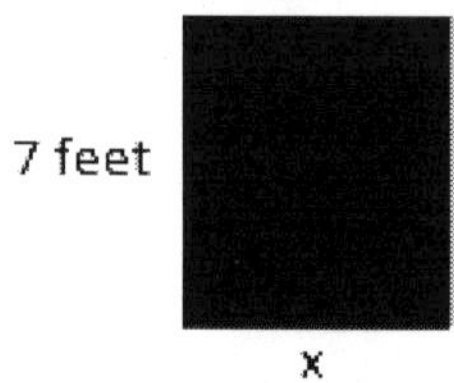

x

Ⓐ **x = 3 feet**
Ⓑ **x = 6 feet**
Ⓒ **x = 13 feet**
Ⓓ **x = 7 feet**

22. **The city is building a fence around a park. The park is 20 feet by 50 feet. If they only want the fence on 3 sides, what is the least amount of fencing they could buy?**

Ⓐ **140 feet**
Ⓑ **100 feet**
Ⓒ **120 feet**
Ⓓ **90 feet**

23. The perimeter of the following object is 38 feet. Find the length of the missing side.

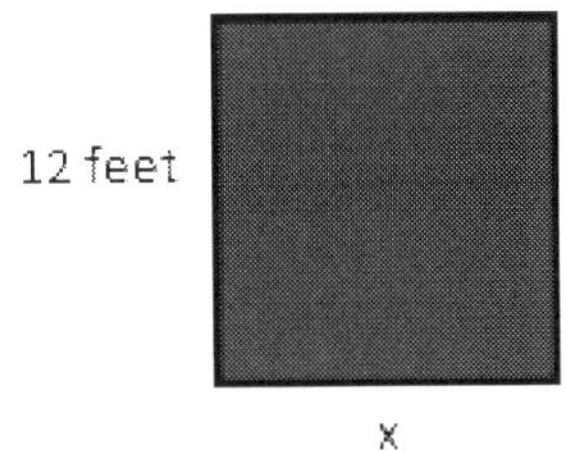

Ⓐ x = 17 feet
Ⓑ x = 12 feet
Ⓒ x = 7 feet
Ⓓ x = 26 feet

24. The perimeter of the following object is 24 inches. Find the length of the missing side.

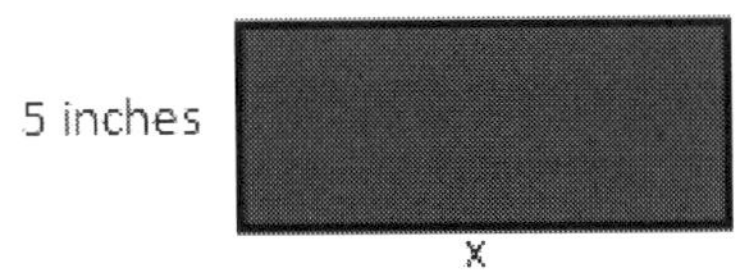

Ⓐ x = 19 inches
Ⓑ x = 7 inches
Ⓒ x = 9 inches
Ⓓ x = 12 inches

25. The perimeter of the following object is 54 yards. Find the length of the missing side.

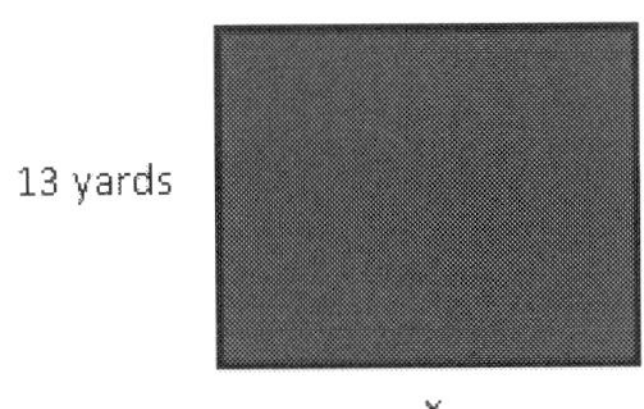

Ⓐ x = 28 yards
Ⓑ x = 14 yards
Ⓒ x = 41 yards
Ⓓ x = 27 yards

26. Which of the following statements are true? Select all correct answers.

Ⓐ The perimeter is the distance around the outside of a plane figure.
Ⓑ The perimeter is the center of a circle.
Ⓒ The perimeter can be found by adding the angles of a triangle.
Ⓓ The perimeter can be found by adding the length of a figure's sides.

27. What is the perimeter of the rectangle shown in figure below? Write your answer in the box given

3 units

6 units

28. Circle the rhombus that has a perimeter of 8.0 cm.

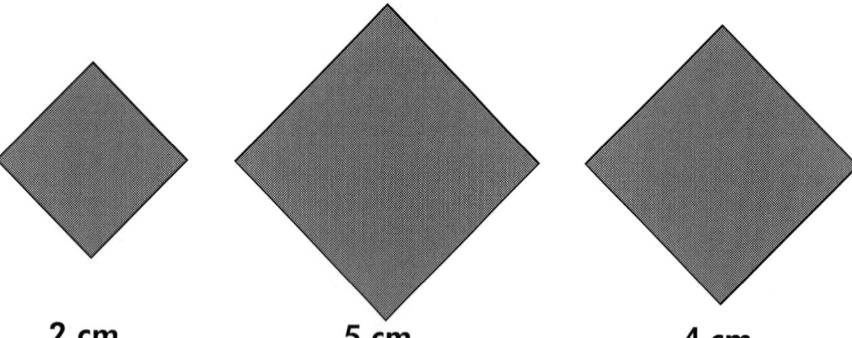

2 cm 5 cm 4 cm

29. John draws a regular hexagon. Each side measures 8 centimeters. He also draws a regular octagon. Each side of the octagon measures 7 centimeters. Which shape has a greater perimeter? How did you arrive at the answer?

30. The perimeters of the rectangles are given in the first column. For each perimeter, select the possible areas of the rectangles.
Note that for each perimeter, more than one option may be correct.
Instruction: Assume that the length and the width of the rectangles are whole numbers.

	15 sq. cm.	10 sq. cm.	12 sq. cm.
Perimeter = 16 cm	○	○	○
Perimeter = 14 cm	○	○	○
Perimeter = 22 cm	○	○	○

End of Measurement & Data

Chapter 5

Geometry

Test ID	3M024
Test Name	2-Dimensional Shapes
Student Name	
Date	

1	Ⓐ Ⓑ Ⓒ Ⓓ	6	Ⓐ Ⓑ Ⓒ Ⓓ	11	Ⓐ Ⓑ Ⓒ Ⓓ	16	Answer in the space provided below.
2	Ⓐ Ⓑ Ⓒ Ⓓ	7	Ⓐ Ⓑ Ⓒ Ⓓ	12	Ⓐ Ⓑ Ⓒ Ⓓ	17	Answer in the space provided below.
3	Ⓐ Ⓑ Ⓒ Ⓓ	8	Ⓐ Ⓑ Ⓒ Ⓓ	13	Ⓐ Ⓑ Ⓒ Ⓓ	18	Answer in the space provided below.
4	Ⓐ Ⓑ Ⓒ Ⓓ	9	Ⓐ Ⓑ Ⓒ Ⓓ	14	Ⓐ Ⓑ Ⓒ Ⓓ	19	Answer in the space provided below.
5	Ⓐ Ⓑ Ⓒ Ⓓ	10	Ⓐ Ⓑ Ⓒ Ⓓ	15	Ⓐ Ⓑ Ⓒ Ⓓ	20	Ⓐ Ⓑ Ⓒ Ⓓ

16

	Yes	No
Circle		
Star		
Square		
Rectangle		

17

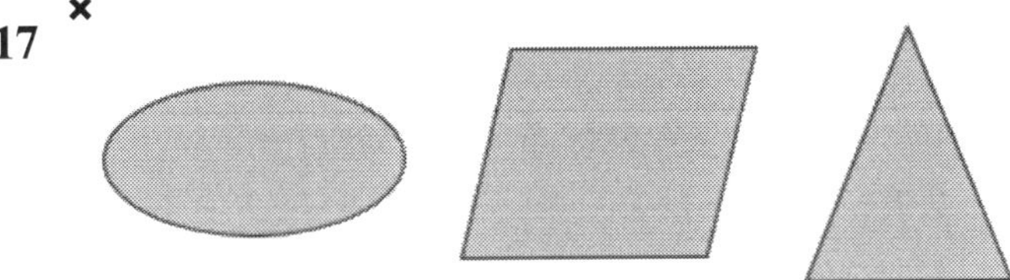

18

Polygon	Attribute	True or False
Rhombus	It has two sets of parallel sides	True
Parallelogram	All the angles are equal	
Rectangle	Opposite sides are equal	

19

Date of Completion:_______________ Score:_______________

Lesson 1: 2-Dimensional Shapes

1. Fill in the blank with the correct term.
Closed, plane figures that have straight sides are called _________ .

Ⓐ parallelograms
Ⓑ line segments
Ⓒ polygons
Ⓓ squares

2. Which of the following shapes is not a polygon?

Ⓐ Square
Ⓑ Hexagon
Ⓒ Circle
Ⓓ Pentagon

3. Complete this statement.
A rectangle must have _____________ .

Ⓐ four right angles
Ⓑ four straight angles
Ⓒ four obtuse angles
Ⓓ four acute angles

4. How many sides does a trapezoid have?

Ⓐ 4
Ⓑ 8
Ⓒ 6
Ⓓ 10

5. Complete the following statement.
A square is always a ________ .

Ⓐ rhombus
Ⓑ parallelogram
Ⓒ quadrilateral
Ⓓ All of the above

6. Which of these statements is true?

Ⓐ A square and a triangle have the same number of angles.
Ⓑ A triangle has more angles than a square.
Ⓒ A square has more angles than a triangle.
Ⓓ A square and a triangle each have no angles.

7. Which of these statements is true?

Ⓐ A rectangle has more sides than a trapezoid.
Ⓑ A parallelogram and a trapezoid have the same number of sides.
Ⓒ A triangle has more sides than a trapezoid.
Ⓓ A triangle has more sides than a square.

8. Complete this statement.
A trapezoid must have ____________.

Ⓐ two acute angles
Ⓑ two right angles
Ⓒ one pair of parallel sides
Ⓓ two pairs of parallel sides

9. Complete the following statement.
Squares, rectangles, rhombi and trapezoids are all _________.

Ⓐ triangles
Ⓑ quadrilaterals
Ⓒ angles
Ⓓ round

10. Which of these shapes is a quadrilateral?

Ⓐ circle
Ⓑ triangle
Ⓒ rectangle
Ⓓ pentagon

11. Which of these shapes is NOT a quadrilateral?

Ⓐ square
Ⓑ trapezoid
Ⓒ rectangle
Ⓓ triangle

12. Name the figure shown below.

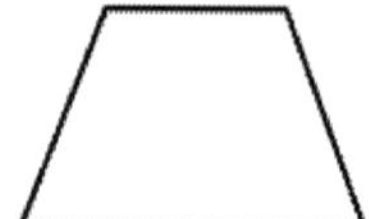

Ⓐ Trapezoid
Ⓑ Square
Ⓒ Pentagon
Ⓓ Rhombus

13. Name the object shown below.

Ⓐ Rectangle
Ⓑ Parallelogram
Ⓒ Trapezoid
Ⓓ Rhombus

14. The figure shown below is a ___________ .

Ⓐ parallelogram
Ⓑ rectangle
Ⓒ quadrilateral
Ⓓ All of the above

15. The figure below is a ___________ .

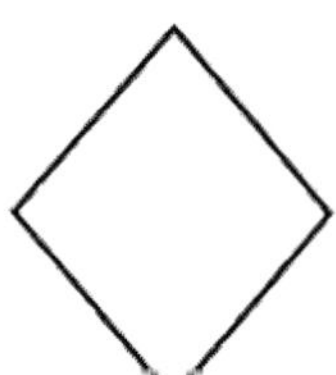

Ⓐ triangle
Ⓑ square
Ⓒ rhombus
Ⓓ trapezoid

16. Are these figures quadrilaterals? Select yes or no.

	Yes	No
Circle		
Star		
Square		
Rectangle		

17. Circle the parallelogram.

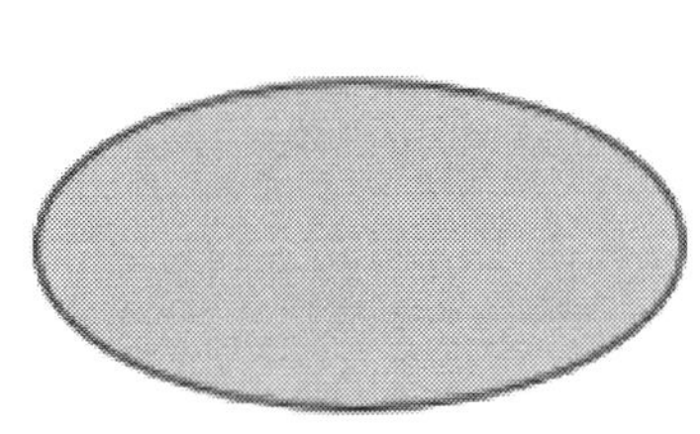
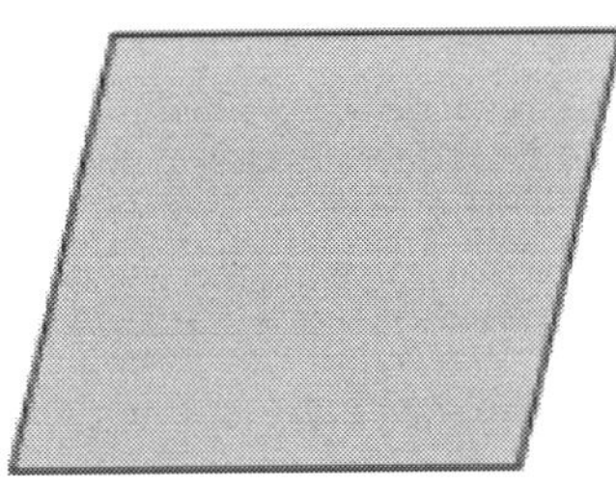
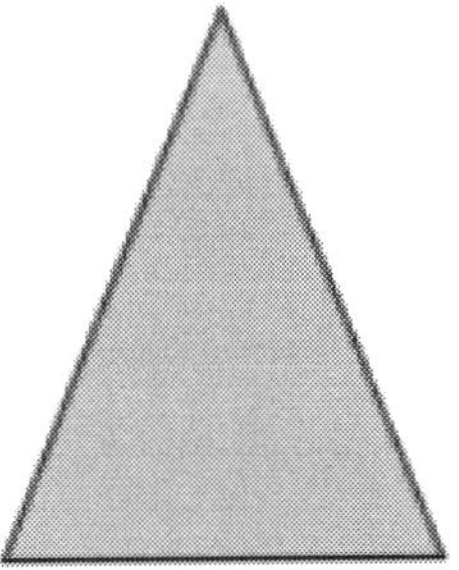

18. For each polygon in the first column, an attribute is defined in the second column. Write true, if the polygon has the mentioned attribute or write false if the polygon does not have the mentioned attribute.

Polygon	Attribute	True or False
Rhombus	It has two sets of parallel sides	True
Parallelogram	All the angles are equal	
Rectangle	Opposite sides are equal	

19. **Draw a quadrilateral which has three obtuse angles.**
Instruction : An obtuse angle is an angle which measures more than 90° but less than 180°.

20. **Which of the following figures have at least one set parallel sides? Note that more than one option may be correct.**

Ⓐ 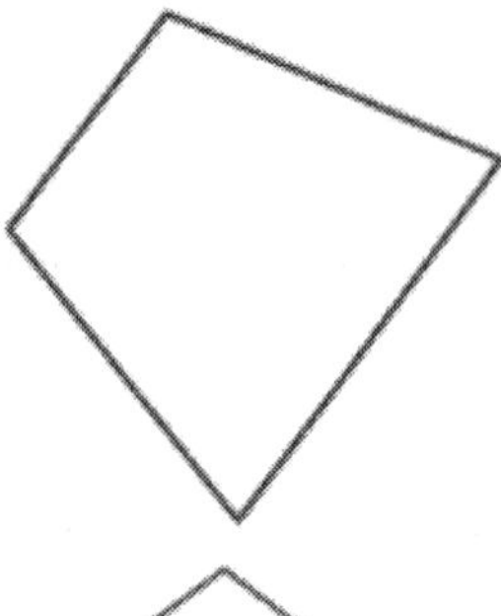

Ⓑ 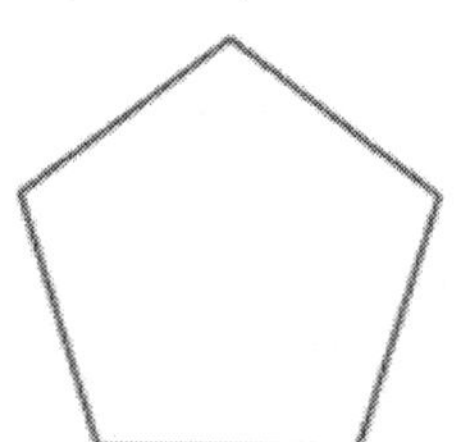

Ⓒ 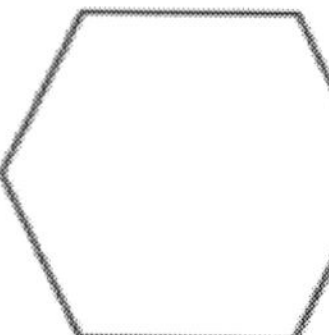

Ⓓ 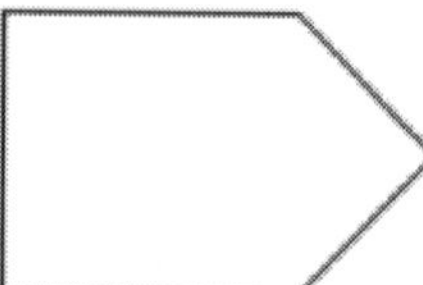

Test ID	3M025
Test Name	Shape Partitions
Student Name	
Date	

1	Ⓐ Ⓑ Ⓒ Ⓓ	6	Ⓐ Ⓑ Ⓒ Ⓓ	11	Ⓐ Ⓑ Ⓒ Ⓓ	16	Answer in the space provided below.
2	Ⓐ Ⓑ Ⓒ Ⓓ	7	Ⓐ Ⓑ Ⓒ Ⓓ	12	Ⓐ Ⓑ Ⓒ Ⓓ	17	Answer in the space provided below.
3	Ⓐ Ⓑ Ⓒ Ⓓ	8	Ⓐ Ⓑ Ⓒ Ⓓ	13	Ⓐ Ⓑ Ⓒ Ⓓ	18	Answer in the space provided below.
4	Ⓐ Ⓑ Ⓒ Ⓓ	9	Ⓐ Ⓑ Ⓒ Ⓓ	14	Ⓐ Ⓑ Ⓒ Ⓓ	19	Answer in the space provided below.
5	Ⓐ Ⓑ Ⓒ Ⓓ	10	Ⓐ Ⓑ Ⓒ Ⓓ	15	Ⓐ Ⓑ Ⓒ Ⓓ	20	Ⓐ Ⓑ Ⓒ Ⓓ

16

	Yes	No

17

18

19

Date of Completion:_______________ Score:_______________

Chapter 5 → Lesson 2: Shape Partitions

1. What is the dotted line that divides a shape into two equal parts called?

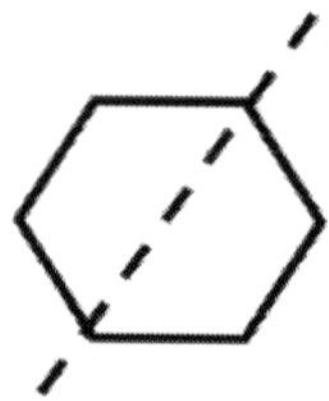

Ⓐ a middle line
Ⓑ a line of symmetry
Ⓒ a line of congruency
Ⓓ a divider

2. A square has how many lines of symmetry?

Ⓐ 8
Ⓑ 4
Ⓒ 1
Ⓓ 2

3. Which of the following has NO lines of symmetry?

Ⓐ 

Ⓑ 

Ⓒ 

Ⓓ 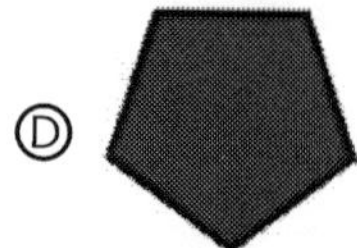

4. Which of the following objects has more than one line of symmetry?

Ⓐ 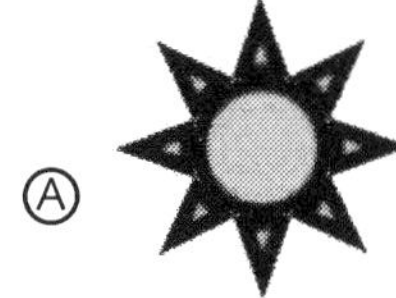

Ⓑ 

Ⓒ 

Ⓓ 

5. What fraction of this triangle is shaded?

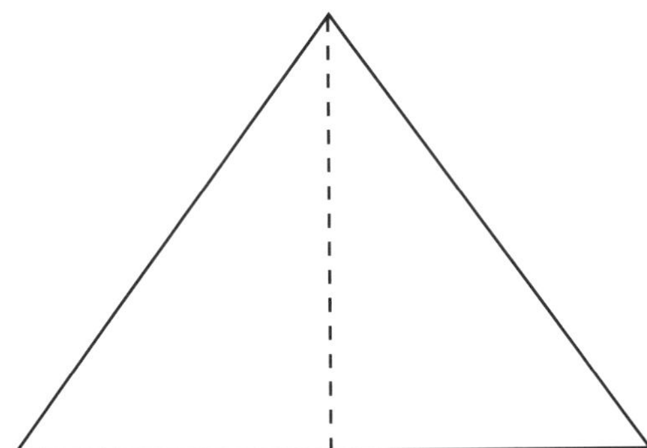

Ⓐ $\frac{1}{2}$

Ⓑ $\frac{2}{2}$

Ⓒ $\frac{0}{2}$

Ⓓ $\frac{3}{4}$

6. **What fraction of this triangle is shaded?**

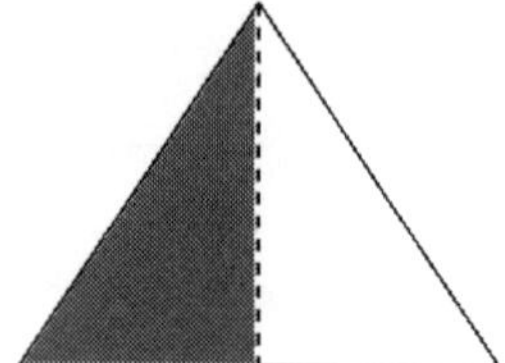

Ⓐ $\frac{0}{2}$

Ⓑ $\frac{1}{2}$

Ⓒ $\frac{2}{2}$

Ⓓ $\frac{3}{4}$

7. **What fraction of this triangle is shaded?**

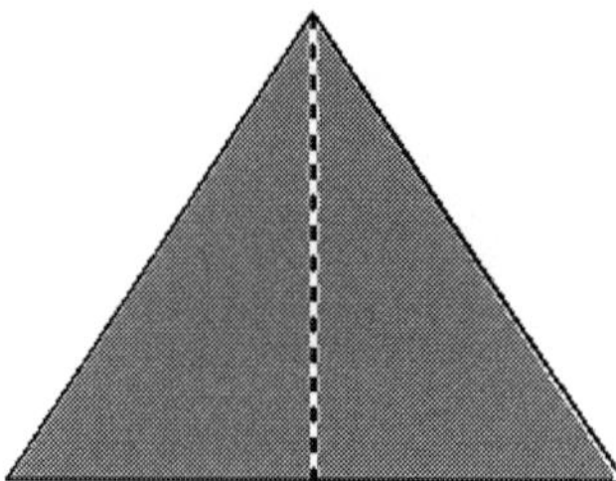

Ⓐ $\frac{0}{2}$

Ⓑ $\frac{1}{2}$

Ⓒ $\frac{2}{2}$

Ⓓ $\frac{3}{4}$

8. What fraction of this square is shaded?

Ⓐ $\frac{0}{2}$

Ⓑ $\frac{1}{2}$

Ⓒ $\frac{2}{2}$

Ⓓ $\frac{3}{4}$

9. What fraction of this square is shaded?

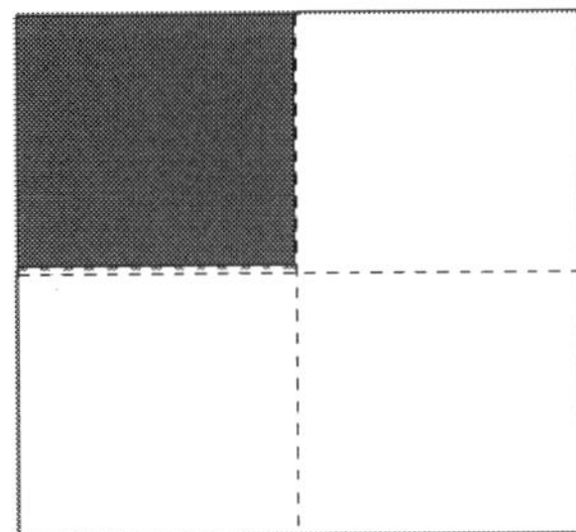

Ⓐ $\frac{0}{4}$

Ⓑ $\frac{1}{4}$

Ⓒ $\frac{2}{4}$

Ⓓ $\frac{1}{2}$

10. What fraction of this square is shaded?

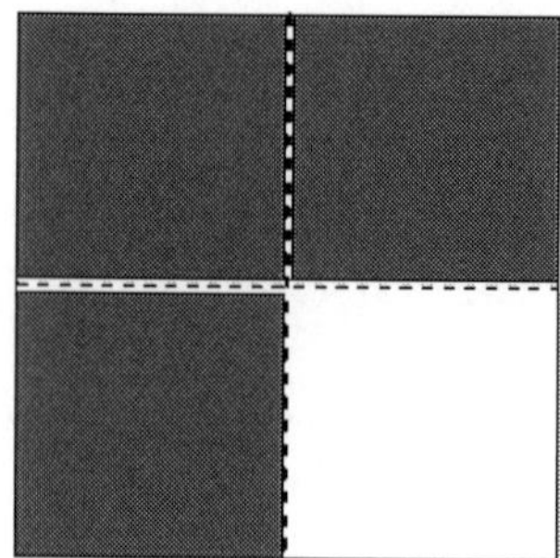

Ⓐ $\frac{0}{4}$

Ⓑ $\frac{1}{4}$

Ⓒ $\frac{1}{2}$

Ⓓ $\frac{3}{4}$

11. What fraction of this rectangle is shaded?

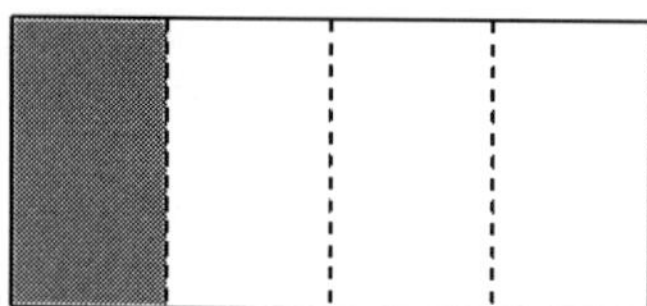

Ⓐ $\frac{0}{4}$

Ⓑ $\frac{1}{4}$

Ⓒ $\frac{2}{4}$

Ⓓ $\frac{3}{4}$

12. What fraction of this circle is shaded?

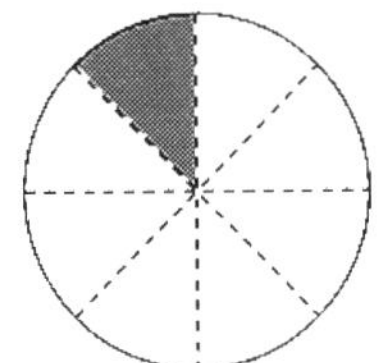

Ⓐ $\frac{1}{8}$

Ⓑ $\frac{1}{4}$

Ⓒ $\frac{1}{2}$

Ⓓ $\frac{3}{4}$

13. What fraction of this circle is shaded?

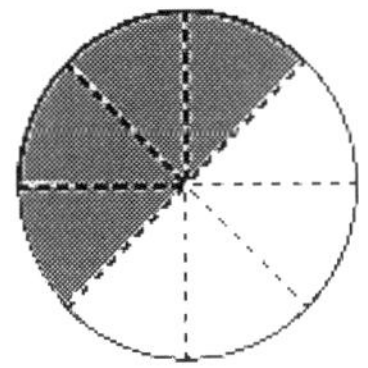

Ⓐ $\frac{1}{8}$

Ⓑ $\frac{1}{4}$

Ⓒ $\frac{5}{8}$

Ⓓ $\frac{4}{8}$

14. The area of the entire rectangle shown below is 48 square feet. What is the area of the shaded portion?

Ⓐ 36 square feet
Ⓑ 48 square feet
Ⓒ 144 square feet
Ⓓ 12 square feet

15. If the area of the entire rectangle below is 36 square feet. What is the area of the shaded portion?

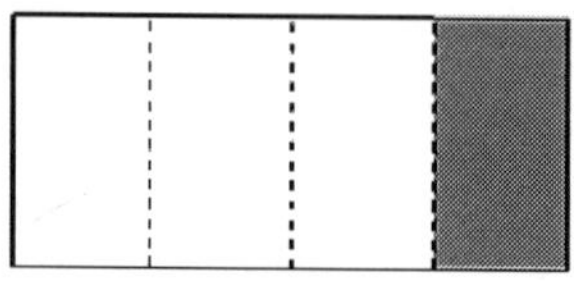

Ⓐ 8 square feet
Ⓑ 9 square feet
Ⓒ 144 square feet
Ⓓ 12 square feet

16. Do these figures have a line of symmetry? Select yes or no.

	Yes	No

17. Circle the shape that has a line of symmetry.

18. If the area of the whole rectangle is 28 square units, what is the area of the shaded portion? Write your answer in the box given below.

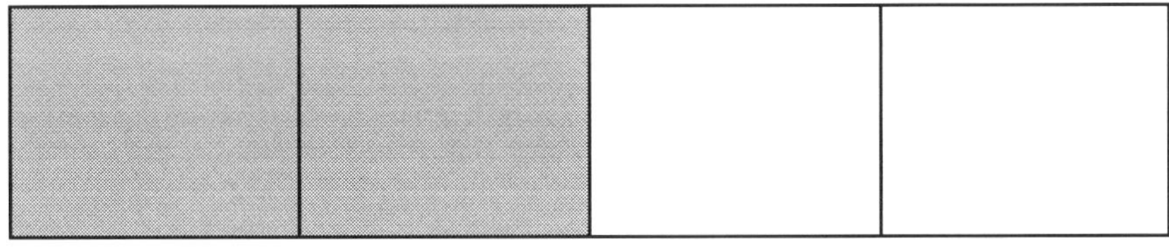

square units

19. Shade one third of the figure below.

20. A circle has an area of 96 sq. cm. The circle is divided into 8 equal parts. Which of the following statements are correct? Select all the correct answers.

Ⓐ If you shade 3 parts, the area of the shaded portion is 32 sq. cm.
Ⓑ If you shade 4 parts, the area of the shaded portion is 48 sq. cm.
Ⓒ If you shade 7 parts, the area of the shaded portion is 84 sq. cm.
Ⓓ If you shade 2 parts, the area of the shaded portion is 24 sq. cm.

End of Geometry

Test Mastery tedBook by Lumos Learning - 3rd Grade Math State Test Prep Workbook | Two Online Grade 3 Math Practice Tests & Learning Resources: Covers Multiplication, Division, Fractions, Addition, Subtraction, Geometry, Measurement (Ages 8-9)

Contributing Author - Leigh Hargett
Contributing Author - Anneda Nettleton
Contributing Author - George Smith
Contributing Author - Wendy Bundgaard
Executive Producer - Mukunda Krishnaswamy
Program Director - Anirudh Agarwal
Designer and Illustrator - Sowmya R.

ISBN 13: 978-1959697497

Printed in the United States of America

CONTACT INFORMATION

LUMOS INFORMATION SERVICES, LLC

 PO Box 1575, Piscataway, NJ 08855-1575
 www.LumosLearning.com

 Email: support@lumoslearning.com
 Tel: (732) 384-0146
Fax: (866) 283-6471

How to Use the Answer Booklet

1. Use the **Table of Contents** to find the lesson you wish to practice.

2. Use a **2B pencil** to answer the questions.
   Marking Instruction: Completely fill in the appropriate bubble.

   Correct ● **Incorrect** ⊗ ✓ ◉ ⊙

3. For open-ended questions, write your answer in the boxes provided on the answer sheet.

4. After completing a lesson, take a clear picture of the answer sheet and upload it using the **SCAN ANSWER SHEET** option provided in the student account. Alternatively, your teacher may also scan and upload the answer sheet.

5. Once uploaded, follow the instructions to get the results. Open-ended questions will be graded by your teacher.

Made in the USA
Middletown, DE
20 March 2024